Nine To

Nine to the Ninth Power of Nine:

Supreme Mathematic African Ma'at Magic

Version 1.0

Written By

African Creation Energy

www.AfricanCreationEnergy.com

September 9, 2009

African Liberation

Science, Mathematics, and Technology

(S.M.A.T.)

Book 2

Supreme Mathematic African Ma'at Magic

All Rights Reserved
Copyright © 2010 by African Creation Energy,
Written and Illustrated by African Creation Energy,
www.AfricanCreationEnergy.com
No part of this book may be reproduced or transmitted in any form or by any means, electronic or mechanical, including photocopying, recording, or by any information storage and retrieval system without permission in writing from the author.

ISBN 978-0-557-59214-2

Printed in the United States of America

Nine To The Ninth Power of Nine

AFRICAN CREATION ENERGY ALERT:

In chemistry, "**Activation Energy**" is defined as the energy that must be overcome in order for a chemical reaction to occur. A "Chemical Reaction" is a process that leads to the transformation of one set of chemical substances to another. Chemical reactions can be either spontaneous (requiring no input of energy) or non-spontaneous (requiring the input of some form of energy). A "Chemical Weapon" is the use of "chemical substances" (material with specific properties) that can cause disturbances to organisms usually by chemical reaction when a sufficient quantity is absorbed by an organism. The symbol above is a modification of the warning symbol for a chemical weapon. In order for the Human brain to understand or comprehend information or input received, Chemical reactions take place in the brain that transform the information into a form that is able to be comprehended and usable. The energy that must be overcome before the chemical reaction of Comprehension and Understanding can take place is mental "Activation Energy". This book deals with the subject of Mathematics which is a subject that is not easily or readily understood or comprehended by most people, however, the subject of Mathematics is extremely important when it comes to science, engineering, problem solving, architecture, invention, innovation, and creating. Therefore, African Creation Energy's modified version of the "Chemical Weapon" warning symbol appears here as a sign to signify that the information contained in this book is intended to be a metaphorical chemical weapon in the sense that the intent of this book is to efficiently present information and transmit motivational energy which should serve as mental Activation Energy to encourage the mental Chemical Reaction process of understanding and comprehension of the subject of Mathematics in the mind of African people which can then be used to create whatever is needed or desired. **Proceeded if you Will.**

African Creation Energy

Creative Solution-Based Technical Consulting

Table of Contents

Section	Page
1. What is Mathematics..	05
2. African Origin of Mathematics....................................	20
3. African Number Systems..	38
4. The Math of Ma'at: Logic and Order...........................	53
5. The Math of Myth: Probability....................................	70
6. The Myth of Math: Numerology..................................	79
7. Sacred Geometry Mathematic Objects........................	102
8. The Math of Math and the Math in Math....................	112
9. Nine To The Ninth Power of Nine...............................	119
0. About the Author: What is African Creation Energy?...	128
References, Resources, and Photo Credits...................	142

1. WHAT IS MATHEMATICS?

Mathematics is an intellectual and scientific tool created and invented by African people for the purpose of systematically studying, learning, and comprehending **Nature**. Since Mathematics is the intellectual tool used for studying Nature, there is a field of Mathematics that corresponds to every aspect of Nature. In the broadest sense, Nature consists of Space/Vacuum, Matter/Energy, and Time/Existence. The field of Mathematics called **Arithmetic** deals with basic counting or quantifying things found in nature with similar qualities. **Algebra** studies the formulas and equations of Nature, **Geometry** and **Trigonometry** studies the Spaces, Shapes and Forms of Nature, **Calculus** studies the changes over time in Nature. Even the mental faculties of Humans are studied in fields such as **Mathematical Logic** using **Binary Algebra**, which studies **Logic and Reason**, and **Statistics** which studies **speculation and probability (belief)**.

Formal definitions of Mathematics state that Mathematics is the Scientific and systematic treatment and study of Quantity, Magnitude, Multitude, Form, Structure, Space, and Change using symbolic objects to discover patterns, find solutions, determine conclusions, and **establish truth** using rigorous **deductive reasoning**. The etymology of the word "Mathematics" comes from the Greek word *"Mathema-"*

meaning "to Learn or to think; to have one's mind aroused; to be wise, **wide awake** and lively". The suffix *"-tic"* (shortened version of the Greek word *"tekhne"* as in "Technique") in the word "Mathematics", means "system, process, method, art, or craft". Thus, the etymological sense of the word "Mathematics" is "The art of Thinking," "Systematic Learning," or "**Thought Process**".

Modern Scientist classify Mathematics as a "Formal Science" to distinguish it from other fields of Science called "Natural Sciences" (Physics, Chemistry, Biology, etc) because the Natural Sciences rely on Experimentation, Evidence, and **The Scientific Method** to arrive at scientific results and conclusions, whereas Mathematicians use definitions, abstractions, logic, and reason as the basis of mathematical proofs, concepts, and conclusions. The four basic steps to "The Scientific Method" are 1) Observation, 2) Hypothesis, 3) Testing, and 4) Conclusions, while the four basic steps to "The Mathematical Method" are 1) Comprehension, 2) Analysis, 3) Synthesis, and 4) Conclusions. The mental ideas, concepts, symbols, and abstractions that form the basis of Mathematics are called "Mathematical Objects". Mathematical Objects are used to model and represent various observed aspects in reality and the natural world. Examples of some Modern Mathematical Objects include **Numbers**, Functions, formulas, equations, groups, sets, graphs, matrices, points, lines, triangles, circles, squares, cubes, etc, Mathematic objects exist as concepts and do not exist in reality, however,

Mathematic objects provide a way for the human mind to comprehend, explain, and model reality. Proper understanding of this point enables the observer to distinguish what exists as a concept and what exists in reality. For example, it cannot be proven that the number 2 exists in reality, however, it can be proven that based on the definition of what the number 2 represents, examples can be found in reality that correspond to what the number 2 represents by definition - The same is true for all Mathematical Objects. Thus, the best and most efficient Mathematical Objects are those concepts and ideas whose symbols closely correspond to the real phenomenon that they intend to represent. This mental process of conceptually, figuratively, and symbolically depicting, portraying, and displaying real natural phenomenon for the purpose of learning, comprehending, applying, and manipulating Nature began with the earliest humans in Africa. And so, some examples of the earliest <u>African Mathematical Objects</u> include many of the gods and goddesses like Ptah, Re, Tehuti, Ma'at, and Heka and their associated signs and symbols.

Supreme Mathematic African Ma'at Magic

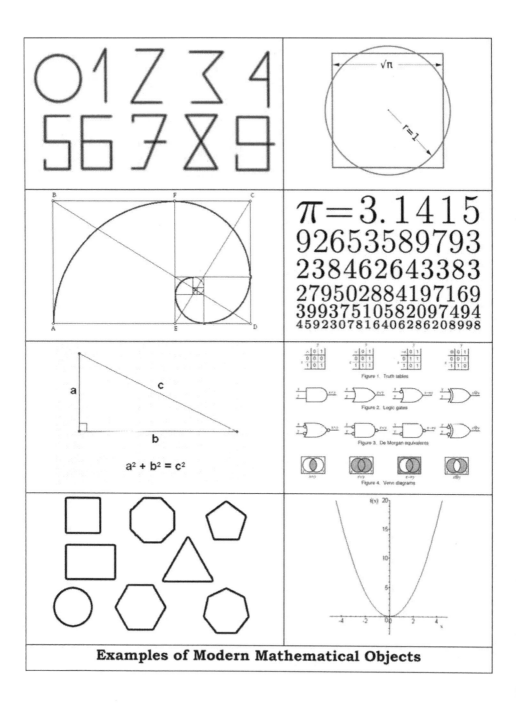

Examples of Modern Mathematical Objects

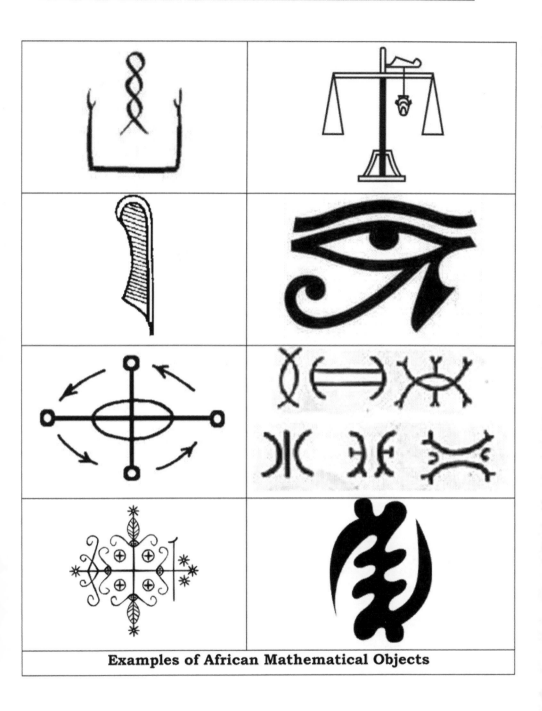

Examples of African Mathematical Objects

Mathematics is the Science that can be found in all fields of Science, thus, **Mathematics is THE SCIENCE in Sciences**. It is the combination of Mathematics based on logic and Reason with Natural Sciences based on The Evidence and the Scientific Method that enables correct and accurate descriptions, comprehension, and explanation of Nature and Reality. Mathematical models that accurately describe the evidence experienced in the Natural Sciences facilitates the ability to model, replicate, and recreate the various observed natural phenomenon. Mathematical knowledge cannot be applied in reality without studying the natural sciences, and conversely, the Knowledge obtained using evidence and experience in the Natural Sciences cannot be applied without Mathematics. The purpose of any type of Knowledge is to <u>use and apply it</u>. It is the union of the evidence and experience from the Natural Sciences with the Logic and Reason from Mathematics that makes it possible to use and apply any Knowledge obtained for survival and well-being. Mathematics is a mental process of finding and figuring things out. Mathematics is the mental process of finding the answer to a question or the solution to a problem. Mathematics is the solution form of Reason. (Or solution reason) and is the mind concentrating upon finding the answer to a question of the solution to a problem. Mathematics as the solution form of reason is Naturally Solvent (capable of solving problems).

The ability to replicate, model, manipulate, and control various aspects of Nature using Mathematics was known to the earliest Africans as **Heka**. Heka as a concept was used to represent the manipulation of the elements of Nature. Heka is often defined as "Magic". **Magic** is defined as "*the practice of producing a desired effect or result by techniques that enable humans to control forces of nature*". In modern terms, Heka and Magic would be called "Engineering". **Engineering** is defined as "*the process of acquiring and applying scientific and mathematical knowledge to design and develop inventions that realize a desired objective*". When looking at the definitions of Engineering, Magic, and Heka the relationship can be seen, and successfully practicing Engineering, Heka, or Magic depends on mathematics.

It is interesting to note that the etymology of the word Engineer comes from the word "engin" meaning "skill and cleverness" and also the word "ingenium" meaning "**inborn qualities** or talent" and "**gene**" meaning "**to beget or produce**". Given the etymology of the word Engineer, and the definition of the word Engineer as someone who practices Heka or Magic (the process of acquiring and applying scientific and mathematic knowledge for the purpose of producing a desired effect), it implies that this Engineering ability is related to genes, genetics, or certain "inborn qualities". It is also true that the ancient Africans viewed the ability to practice Magic or Heka (Engineering) with certain blood lineages or genetics.

In Mathematics, there is a field called "**Pure Mathematics**," also called "**Speculative Mathematics**," which is performing mathematics solely for the purpose of "displaying eloquent logic and reason" (ritual) without the intention of ever applying the mathematical concepts to the real world. This is the opposite of the field of Mathematics called "**Applied Mathematics**" also called "**Operative Mathematics**". Applied Mathematics is the field of mathematics dedicated to the application and utilization of mathematics in all other sciences and disciplines. The peculiar reality about "Pure Math" or "Speculative" math is that while it may be performed without the intention of being applied, if the Mathematical concepts developed are based on rigorous **deductive reasoning**, then an application in the real world for the abstract mathematical concept is usually discovered later. This fact shows that the real world is based on **Reason**. The danger of "Pure math" is if rigorous deductive reasoning is abandoned, degraded, disregarded, or forgotten while performing "Pure Math" or "Speculative Mathematic" thought processes of any type, then it gives way to creating mental objects that have no application in the real world that are literally useless and serve the purpose of being only "mind numbing" entertainment. "Pure Math" or Speculative Mathematics without rigorous deductive reasoning gives way to the esoteric speculative field of Numerology.

Numerology by definition is any system or set of ideas, beliefs, or hypothesis about the relationship between Numbers and reality. There is nothing wrong or incorrect about forming hypothesis and beliefs, however, when conclusions are drawn about the hypothesis and beliefs without appropriately testing them using The Scientific Method in the case of the Natural Sciences, or without rigorous deductive reasoning in the case of Mathematics and the formal sciences, then it leads to inaccurate explanations of reality that cannot be used and applied, and wasted time, energy, and effort of individuals who have accepted the inaccurate explanations as truth and have attempted to apply them. Because Numerological conclusions and explanations are based on untested, loosely tested, or poorly tested beliefs and hypothesis, it is considered a "pseudo-science" or "pseudo-mathematic" because all of the results from numerological explanations and conclusions are not consistent when tested each time the conditions are the same. However, Numerology by definition, which is a belief about a number's (or mathematical object) relationship to reality, is essential to Mathematics, but what separates Mathematics from Numerology is the presence of rigorous Deductive Reasoning and consistency. Many of the earliest mathematicians were Numerologist, and many of the present-day modern Mathematicians are numerologist to some degree. Therefore, in the field of Numerology, there does exist what could be considered "Operative Numerology". **Operative Numerology** would be the conclusions and explanations from the field of Numerology that are consistent, testable, reasonable, and applicable. Numerology gives birth to Mathematics through Reason.

order by Usefulness	Subject	Order by origin
1	Applied (Operative) Mathematics	4
2	Pure (Speculative) Mathematics	3
3	Applied (Operative) Numerology	2
4	Pure (Speculative) Numerology	1

In terms of numerology and mathematics: The word **Math** comes from the Greek word "*metheus*" or "*mathematos*" meaning "**To think** or to learn" and the word **Myth** comes from the Greek word "Mythos" meaning "**Thought**, speech, or story". Both Math and Myth refer to Thought. **Myth** becomes **Math** by way of evidence and reason; this is the forward process of the Human Mind experiencing Nature, speculating on Explanations for the experience, testing the speculation for validity using evidence and reasoning, and if the speculation is proved valid, declaring a Mathematical (or Scientific) principle. **Math** becomes a **Myth** by way of speculation without evidence and reasoning; this is the backward process of the Human mind experiencing a Mathematical principle declared by someone else without knowing the reasons, proofs, or meaning of the Mathematic principle. Numerology is Mathematic Mythology, and can be valid or invalid depending on if the Numerological principle was declared using the forward or backward process of the Human mind.

Let's consider the ancient mathematical problem of "Squaring the circle". "Squaring the Circle" is the process of trying to construct a Square that has the same area as a Circle using only a compass and a straightedge. Mathematics represents the objective, finite, operative, exoteric Quantification science, while Numerology represents the subjective, infinite, speculative, esoteric Quantification science. The principles of the field of Mathematics could be represented by the geometric symbol of the "Square" and the principles of the field of Numerology could be represented by the geometric symbol of the "Circle". Esoterically, in the field of Numerology, the process of "Squaring the Circle" is a metaphor for attempting to make the infinite become finite and make the unknown become known (Making Mathematics equivalent to Numerology for example). Exoterically, in the field of Mathematics, the geometric problem of "squaring the circle" does have a theoretical solution, but the theoretical solution is considered impossible to apply or construct using a compass and a straightedge tool because the equation for the area of a circle contains the infinite, transcendental constant number **Pi** (π=3.1459...). Mathematically, while Squares and Circles are both equal in degree with both shapes having a total of 360 degrees each, they are different in their formulas for perimeter (circumference) and area.

Alike or Different?
Alike and Equal in degree, but
Different and Unequal in area and perimeter

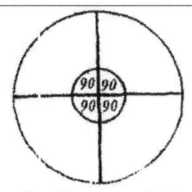

Total Degrees = 360°
Perimeter = π * Diameter
Area = π * radius²

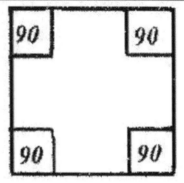

Total Degrees = 360°
Perimeter = 4 * length
Area = length²

"Squaring the Circle"

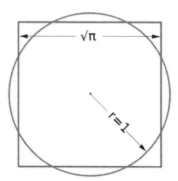

The theoretical solution to the problem is to construct at circle with Radius = 1 (diameter = 2) and a Square with Length = √π (≈1.77245...) where the Area of the Circle and Square would both be Pi (π). The closest whole number approximation that yields a Circle and a Square with similar areas is a circle with a Diameter = 9 (radius = 4.5) and a Square with Length = 8, where the area of the Circle would be approximately 63.6 and the area of the Square would be 64.

As stated earlier, the goal of this book is to convey knowledge and information that is useable (able to be used) and applicable (able to be applied). For this reason, we attempt to metaphorically "Square the Circle" or take the useable and applicable components of both Mathematics and Numerology since both of these fields are African in origin. The field of mathematics called **"Applied Mathematics"** is emphasized due to it being readily practical, applicable, and useable. It is through Mathematics in conjunction with Science that **Potential** African Creation Energy can be transformed into **Kinetic** or Active African Creation Energy for the purpose of achieving the goal of creating everything needed for the survival and well-being of African people, by African people, and for African people.

This book has been titled "Supreme Mathematics, African Ma'at Magic" as a representation of how the information presented in this book attempts to metaphorically "Square the Circle". The term "Supreme Mathematics" sounds rather objective and scientific, however, as it is discussed later, is rather subjective, esoteric, and speculative. Conversely, "African Ma'at Magic" as a term sounds rather subjective and esoteric, however, as it is discussed later, is very objective and scientific just like the fields of mathematics, science and engineering.

The purpose of this book entitled "Nine to the Ninth Power of Nine, Supreme mathematic, African Ma'at Magic" is as follows:

- To show the fact that Mathematics is of African origin

- To show relationships between African mathematical objects, terms, and concepts to Modern Mathematical Objects, terms, and concepts.

- To show how African Mathematical methods can be used to solve certain problems not easily solved by modern mathematical methods

- To Show the Importance that Ancient Africans placed on Mathematics to **solve All Problems**, establish **Truth and Order (Ma'at)**, and **Apply knowledge** to create, develop, and engineer any and all Systems and Technologies needed for survival and well-being

This book and other books written and published by the coined term "**African Creation Energy**" are introductory educational tools for a long term goal and mission of growing, cultivating, nurturing, encouraging, supporting, advancing, and promoting

African Creativity, Inventiveness, and Ingenuity for the purpose of developing, engineering, forming, formulating, innovating, inventing, designing, building, and creating **any** materials, structures, machines, devices, systems, and processes needed for survival and well-being by African people for African people. With that said, the importance of Mathematics is illustrated, described, and emphasized in this book. However, this book does not teach Mathematics or Mathematical concepts (there are numerous text books that serve that purpose) but rather shares and provides **Motivation**, **inspiration**, and **Insight** into the relationship between Modern Mathematics and Technology to Ancient African mathematics and technology.

2. THE AFRICAN ORIGIN OF MATHEMATICS

The Greek word "Mathema" meaning "thought" in the word "Mathematics" is also related to the word "metheus" meaning "Thought" that can be found in the name of the Greek deity **Prometheus** (meaning "Fore-thought"). In Greek mythology, Prometheus was the son of **Themis** (Greek deity called "**Lady of Justice**"). Prometheus stole **fire** from **Zeus** and gave it to humanity. Prometheus is comparable to another character in Greek mythology called **Hermes** who was considered the inventor of Fire, the messenger of Greek deities, and Greek deity of wisdom, writing, and invention. The "Fire" in the Greek mythology story that Hermes invented and Prometheus stole from Zeus to give to humanity is symbolic of "Knowledge" or "Information" which Humanity acquires by thinking and learning. This "Fire" or "knowledge" could be used in either a **constructive** or **destructive** manner. This fire for all intents and purposes is Mathematics, the human Thought Process.

The same principles found in the "Prometheus steals Fire" story can be found in the Biblical Genesis story of the Serpent (who is equated with **Lucifer** the Light or **Torch Bearer**) in the Garden of Eden who convinced Adam and Eve to eat from the "Tree of Knowledge" which God attempted to keep from humanity. Another story which parallels the Prometheus story about humanity's acquisition of "Fire" (Knowledge) is the story of **Obassi Osaw** amongst the **Efik** people in Niger and Nigeria in Africa. According to the Ekoi's story, Obassi Osaw (also spelled **Abassi**, the Sky God and creator deity) choose not to give fire to the first humans he created. One day, a young boy was able to steal fire from Obassi Osaw and bring it back to humanity.

Both the Biblical story and The Greek story were modeled after the African stories which predated them. The Greek deities Prometheus and Hermes in particular were modeled after the African Egyptian deity Tehuti. **Tehuti** (also called Djehuti, Zehuti, or Thoth) was the Egyptian god of Wisdom, Writing, Magic, Sciences, Math, and measuring time. Tehuti was a lunar deity and often depicted as a man with the head of an Ibis bird or as a Baboon. In the Egyptian mythology, Tehuti was the Husband of **Ma'at**, the Egyptian Goddess of Truth, Justice, Order, and Divine Wisdom. **Tehuti** was responsible for transforming the **creative will** of **Ra** (The Egyptian Sun God) into **action**. Tehuti was associated with the philosophical concept of "the Logos" (meaning word or saying and action or work) which is derived from the word "Lego" meaning **"To Count** or To Speak". Logos also is used to indicate the Logic or Reason, purpose, and Hypothesis for speaking. All of this is to show that **the origin of the word "Mathematics" comes from the African deity Tehuti**.

Just as it has been shown that the word "Mathematics" comes from the African Deity Tehuti, it can be shown that the word "Number" (which is essential to Mathematics) comes from Tehuti's wife and African Goddess **Ma'at**. The etymology of the word "Number" is from the Greek word "nemein" meaning "to divide, to distribute, allot, or deal out." The Greek word "nemein" is the root of the word "**Nemesis.**" Nemesis is a Greek goddess who *distributes* or *deals out* retributive justice, vengeance, and divine retribution as a result of disregarding another Greek goddess called "**Themis**" (lady justice) who was the mother of Prometheus and personification of divine order and law. Both Nemesis and Themis were worshipped by the Greeks in a temple called Rhamnous. The prefix "**Rham**-" in the word "Rhamnous" comes from the Greek word "Rhamata" meaning "words, **utterance**, or **command**", and the suffix "-nous" in the Greek word "Rhamnous" comes from the Greek word "**Nous**" meaning "**Mind** or **intellect**" (The relationship between Nous, mind, intellect, divine utterance, and Ma'at will be discussed in greater detail in chapter 4 – The Math of Ma'at). The concepts that these Greek goddesses Nemesis and Themis represented were taken and modeled after the African Goddess Ma'at.

Tehuti was the African Egyptian diety of Thought, Knowledge, wisdom, writing, sciences, and mathematics. Tehuti is often depicted with the body of a man and the head of an Ibis bird, or as a Baboon, or in a more human-looking form as A'ah-Djehuty. Tehuti's wife was Ma'at.

Ma'at was the Egyptian principle of **Truth**, Justice, **Order**, **Balance**, and **Divine Wisdom**. Ma'at brought the Universe into Order from Chaos at the beginning of creation. Ma'at was represented by an ostrich Feather and was also personified as a goddess wearing the Feather. In the Underworld, called the **Duat** (Egyptian Judgment of the dead) in "The Halls of Ma'at", the Heart is weighted against the Feather of Ma'at on the Scales of Ma'at as the measure that determined if the soul adhered to the **42 Laws of Ma'at** and would be able to crossover to the afterlife or the "Here After". This is where religion gets the concept of Judgment after death. The word "Here" is phonetically similar to the word "Hear, therefore "Here-After" is phonetically the same as "Hear-After". If the Letters in "Hear-After" are rearranged you get "Heart-Feather", i.e. the Heart is weighted against the Feather of Ma'at.

Greek Deity Nemesis	**Greek Deity Themis**

Greek Deity Prometheus	**Greek Deity Hermes**

Ma'at was the African Egyptian principle of truth, justice, order, balance, and divine wisdom. Ma'at was symbolized by a Feather, as well as personified as a Goddess wearing a feather. Ma'at was the wife of Tehuti (Mathematics), and Ma'at is the origin of the word "Number". Numbers form the foundation of Mathematics and are used to determine Order, and Mathematic Logic and Reason determine Truth.

Tehuti's female counterpart was the Egyptian Goddess **Seshat**. Seshat (whose name meant "she who scribes") was the Egyptian Goddess of wisdom, knowledge, writing, architecture, astronomy, astrology, building, **mathematics,** and surveying. Seshat bore the title **"The Enumerator"** and **"Lady of the Builders"**. Seshat was depicted as a woman with a papyrus plant (representing writing since the Egyptians made paper out of the papyrus plant) above her head with a pair of inverted cow's horns formed to indicate an inverted crescent moon. Seshat was also depicted writing on a notched palm stem (representing counting the passage of time) and dressed in a leopard skin (representing the stars and a symbol of eternity). To wear a leopard skin in ancient Africa indicated a relationship to the **builders** and **craftsmen** and the ability to practice **Heka** (Magic, engineering). Seshat was considered a Magician (Engineer) and her symbol represented the source of **intuition** and all **creative ideas**.

Whereas Tehuti represented theoretical and **esoteric** mathematic and scientific principles, Seshat represented applicable and **operative** mathematic and scientific principles. Tehuti represented Mathematic theories and Seshat represent Mathematics in practice. The Egyptian principle of Seshat was present when the surveying and Architectural design of Egyptian pyramids and temples was being done to align with certain celestial bodies in the cosmos. Since Ma'at established the order of the cosmos, it was the responsibility of Seshat to reflect Ma'at when designing buildings to be aligned with the cosmos. The deities **Ma'at, Tehuti, and Seshat** represented a triad of concepts related to **Mathematics, Science, and Engineering** that were acknowledged and used by Ancient Africans to develop and build the many structures along the Nile River in Africa.

Supreme Mathematic African Ma'at Magic

Seshat was the African Egyptian Goddess of wisdom, knowledge, writing, architecture, astronomy, astrology, building, mathematics, and surveying.

Nine To The Ninth Power of Nine

Mathematics is an invention that was created by the human mind in order to comprehend and explain nature and reality. It is said that Necessity is the mother of invention. If Necessity is the mother of invention, then it stands to reason that survival or self preservation is the father of invention. Therefore, mathematics was invented because it was necessary for people to comprehend and explain nature and reality for survival. Thus the origin of Mathematics can be found in the same geographic location as the origin of humanity – in Africa.

The earliest examples of mathematics can be found in Africa. The oldest Mathematical artifact found to date is a 37,000 year old item known as the **Lebombo Bone** which was discovered in the African country known as **Swaziland** (also called **Ngwane**) home of the **Zulu** or **Lulu** people. The Lebombo Bone is the Fibula Bone (lower leg bone) of a **Baboon** which was found marked with 29 notches similar to Calendar sticks and Tally Sticks used by the Bushmen of **Namibia** today. Another Mathematical tool similar to the Lebombo Bone that has been discovered is the 20,000 year old **Ishango Bone**. The Ishango Bone was discovered in Africa at the source of the **Nile River** between the countries of **Uganda** and **Congo**. The Ishango Bone, like the Lebombo Bone, is the fibula of a baboon with a piece of **Quartz crystal** attached on one end, and three columns of asymmetrically grouped tally marks carved along the sides. Mathematical analysis of the Ishango Bone indicates knowledge of Sequential Ordering of **Prime Numbers** and the development of a Numeral System that was a precursor to the **Binary** based multiplication method of the Ancient Egyptians that developed in later years further down the Nile.

Both the Lebombo Bone and the Ishango Bone were used as **Lunar Calendars** as indicated by the groupings of 28 to 30 tally marks followed by a distinctive marker. The word Ishango is phonetically similar to "Chang-O" or "Heng-O", a Lunar Deity in China. Archeological evidence indicates that Counting, Arithmetic and thus Mathematics was developed by **African Women** as a way to measure or Quantify Time and chart their menstrual cycle as it related to the Lunar Cycle. It is interesting to note that the Ishango Bone and the Lebombo bone were both parts of a Baboon and were used for calculating and measuring Lunar cycles of time, and the ancient Egyptian deity Tehuti was a **Lunar Deity** often depicted as a **Baboon** and was said to have been responsible for inventing **mathematics** and **measuring time**. Also, it is interesting that most modern clocks use **Quartz crystals** to regulate the Calculating and Measuring of time and that the Ishango bone used for measuring and calculating time was also found containing a piece of Quartz crystal. It has been shown that the oldest Mathematical objects used for quantifying or measuring time have been found in Africa. Also, the oldest mathematical objects used for quantifying or measuring space and matter were developed by Africans as well. The **cubit** is the first recorded unit of length which was originally based on the forearm length of the African Egyptian

Pharaoh or ruler before being constructed into a ruler instrument for measurement.

ABOVE: AFRICAN EGYPTIAN CUBIT RULER, MATHEMATICAL MEASUREMENT INSTRUMENT.

Not only is Africa the home of the oldest discovered Mathematical artifacts, but also the home of the oldest Mathematical writings and Texts ever found. A 4000 year old mathematical papyrus from Egypt's 11th Dynasty is the oldest mathematical writing found to date. This papyrus contains 25 mathematic equations and calculations to determine Surface Area and Volume of geometric figures like hemispheres and frustums (the part of a pyramid between two parallel planes intersecting it), as well as methods to approximate transcendental numbers like **Pi (π)**. This **11th Dynasty Egyptian Geometry papyrus** is now kept in the Museum of Fine Arts in Moscow, Russia and is commonly known as the **"The Moscow Mathematical Papyrus"**.

Left: African lunar Deity Tehuti, god of <u>Mathematics</u> and <u>measuring time</u>, depicted as a <u>Baboon</u>

Above: Ishango Bone; Fibula Bone (lower leg bone) of a <u>Baboon</u> with a piece of <u>Quartz crystal</u> affixed to the left end used for <u>Mathematics</u>, Calculating lunar cycles, and <u>measuring time</u>

Left: Quartz Clock; Modern mathematical tool which uses a <u>Quartz Crystal</u> Oscilator to regulate the calculating and <u>measuring of time</u>

Several other Egyptian Mathematic texts have been discovered to date. These text contain "Word Problems" that demonstrate the various fields of Mathematics that were known by the ancient Africans in Egypt. The "11th Dynasty Egyptian Geometry Papyrus" and **Ahmose's Mathematical Papyrus** (commonly called the **Rhind Papyrus**) are the two primary sources of information on Egyptian Mathematics. The 18th Dynasty scribe and Mathematician **Ahmose** (also spelled Ahmes, meaning "The **Moon is Born**", another reference to the **lunar** deity **Tehuti**) copied the papyrus from an older lost Middle Kingdom papyrus entitled "**Methods for Knowing All Mysteries and Secrets**". In the first paragraphs of the papyrus, Ahmose states that *"Mathematics is the method to inquire into Nature, and obtain correct and right knowledge of all things, All mysteries, and all secrets"*.

List of Discovered Egyptian Mathematic Text		
Egyptian Text NAME	**APPROX. DATE**	**Mathematical CONTENTS**
The Moscow Mathematical Papyrus	2000 BC	25 equations and calculations to determine Surface Area and Volume
The Akhmim Wooden Tablet	2000 BC	Eye of Horus Binary Fractions
The Egyptian Mathematical Leather Roll	1850 BC	table of 26 decompositions of unit fractions
Kahun Papyrus	1825 BC	six mathematical problems using Egyptian Fractions
The Berlin Papyrus	1800 BC	two algebraic problems of simultaneous equations
The Reisner Papyrus	1800 BC	Geometry and volume calculations
The Rhind papyrus	1650 BC	a 2/n TABLE and 84 Problems related to arithmetic, algebra, geometry, trigonometry, metrics, and economics

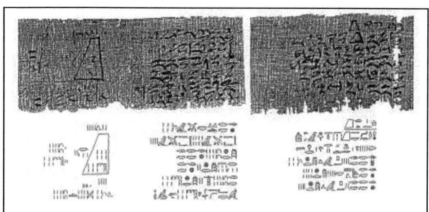

Above: 11th Dynasty Egyptian Geometry Papyrus (The Moscow Mathematical Papyrus)

Left: Kahun Papyrus, 1825 BC, six mathematical problems using Egyptian Fractions

Left: The Reisner Papyrus, 1800 BC, Egyptian Geometry and volume calculations

Supreme Mathematic African Ma'at Magic

Ahmose's Mathematical Papyrus entitled "Methods for Knowing All Mysteries and Secrets" (commonly known as the Rhind Mathematical Papyrus)

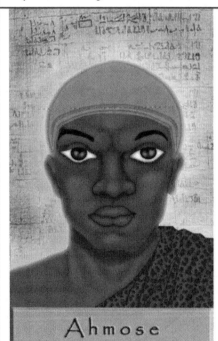

Ahmose, African Egyptian Scribe and Mathematician

Part of Ahmose's Mathematical Papyrus (also known as the Rhind Mathematical Papyrus)

There has been much speculation and fascination for years about the "Secrets of Egypt" and the "Egyptian Mystery System", but according to Ahmose's Mathematical papyrus, the Egyptians viewed mathematics as the means to obtain Right Knowledge about nature, all secrets, and all mysteries. Because it is one of the purposes and intentions of this book to promote efficient use of the energies of African people to create any and all things necessary for survival and well being, a very important point is stressed here: it is encouraged that the same amount of time and energy should be spent learning and applying mathematics that is spent being captivated, fascinated, and spellbound by "mystery systems" and "secret societies", because the Egyptians viewed mathematics as the Key to knowing all mysteries and all secrets. The Ancient Africans in Egypt viewed Mathematics as the key to knowing all mysteries and all secrets because a "secret" or "mystery" in Mathematics is basically an "Unknown" (symbolized in modern Mathematical Algebraic equations and formulas as x), and Mathematics enables the mind to be able to go from one reasonable, sensible, accurate, and correct mental step, idea, or thought to another, eventually solving the problem, answering the question, finding the solution, uncovering the secret, and solving the mystery. Mathematically speaking, Africa is the geometric point of origin from which mathematical concepts exponentially grew, spread, and dispersed throughout the world, and thus all Mathematical concepts can be traced back to the prime source origin of Mathematics which is the African continent and the African Mind.

3. AFRICAN NUMBERING SYSTEMS

A "Number System" is a systematic and methodological way of indicating quantities either as written symbols, spoken words, or some other form of communicating. The Decimal (base-10) Numbering system is the most familiar, widely known, and most commonly used numbering system by most modern people for Mathematical operations. The decimal base-10 Numbering system uses 10 Number symbols (0, 1, 2, 3, 4, 5, 6, 7, 8, and 9) to signify all quantities. In the decimal number system, the symbol for the quantity **9** has the highest value of all single symbols. Any quantity larger than the Number 9 is merely a combination of the 10 Decimal Number symbols. The number of symbols used to indicate a quantity in the Decimal base-10 numbering system depends on powers of 10. For example, quantities less than $10^1=10$ only require 1 symbol, quantities less than $10^2=100$ required 2 symbols, quantities greater than $10^2=100$ but less than $10^3=100$ require 3 symbols, etc. Quantities greater than 0 but less than 1 are expressed as "decimal fractions", and quantities less than 0 follow the same rules as quantities greater than 0, but are displayed with a Negative sign (-) to indicate polarity. Examples of other numbering systems are the Binary Base-2 Numbering system which uses only 2 symbols (0 and 1) to indicate all values, the Octal base-8 Numbering system which uses 8 symbols (0, 1, 2, 3, 4, 5, 6, and 7) to indicate all values, and the Hexadecimal base-16 numbering system which uses 16 symbols (0, 1, 2, 3,

Nine To The Ninth Power of Nine

4, 5, 6, 7, 8, 9, A, B, C, D, E, and F) to indicate all values. In the Binary system, the symbol for the quantity **1** has the highest single value. In the Octal system, the symbol for the quantity **7** has the highest single value, and in the Hexadecimal system, the symbol for the quantity **15** (F) has the highest single value.

\	Various Numbering Systems			
Decimal	**Binary**	**Octal**	**Hexadecimal**	**Vigesimal**
0	00000	0	0	0
1	00001	1	1	1
2	00010	2	2	2
3	00011	3	3	3
4	00100	4	4	4
5	00101	5	5	5
6	00110	6	6	6
7	00111	7	7	7
8	01000	10	8	8
9	01001	11	9	9
10	01010	12	A	A
11	01011	13	B	B
12	01100	14	C	C
13	01101	15	D	D
14	01110	16	E	E
15	01111	17	F	F
16	10000	20	10	G
17	10001	21	11	H
18	10010	22	12	I
19	10011	23	13	J

The most probable origin of the Decimal Number system was the use of the ten fingers on the hands for basic counting and arithmetic functions. In fact the word "Digit" which is used to indicate a single numerical symbol in the decimal Base-10 Numbering system comes from the Latin word "digita" meaning "fingers". While the decimal, base-10 numbering system was and is one of many numbering systems used by Africans, the system of notation used with decimal numbers in mathematics today is a Hindu-Arabic system of decimal number notation.

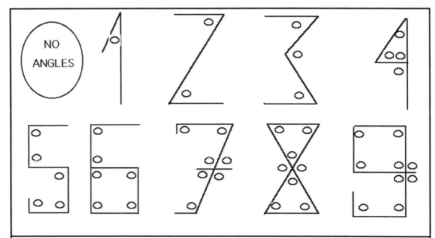

Above: The Hindu-Arabic system of decimal number notation pictured above is the origin of modern English number notation. The value of the number symbol is equivalent to the number of angles in the symbol: for example, the number 1 has only 1 angle, 2 has 2 angles, etc.

Nine To The Ninth Power of Nine

African Decimal number notation is a "tally mark" extension from binary number notation.

Hindu-Arabic Decimal Numbers	African Decimal Numbers							
	Tally Mark Variants		Yoruba	Igbo	Mande Malinke	Efik	Akan (Twi)	Chokwe
0	◇	⊙	Odo					
1	|	|	Eni	Otu	kiling	kyèt	ekó	káxi
2	∧	||	Eji	Abou	fula	ìba	abien'	kári
3	人	|||	Eta	Ato	saba	ìtá	abiêsá	tátu
4	+	|| ||	Erin	Ano	nani	ìnang	anán	uana
5	✕	||| ||	Arun	Ise	lulu	ìtyôn	anúm	mu-tânu
6	✳	||| |||	Efa	Isii	woro	ìtyôkyet	asiá	mu-sambânu
7	✸	|||| |||	Eje	Asaa	worõ-ula	ìtyâba	asón	ximbiári
8	✹	|||| ||||	Ejo	Asato	segi	ìtyâitá	awotwe	nake
9	✺	||| ||| |||	Esan	Itolu	konõto	ùsúk-kyèt	akrón	ívua

Supreme Mathematic African Ma'at Magic

Powers of 10	Hindu-Arabic Decimal Powers of 10	African Decimal Powers of 10
10^0	1	𐤉
10^1	10	∩
10^2	100	ര
10^3	1,000	
10^4	10,000	
10^5	100,000	
10^6 Or Infinity (∞)	1,000,000 Or Infinity (∞)	

As the intellectual tool used for studying, learning, comprehending, and modeling nature, it makes sense that any Numbering Systems that would be used as the fuel to any and all Mathematical operations would also be based on Nature. Therefore, the numbering system that was originally developed by Africans to study and observe nature was the Binary, Base-2 Numbering System. The Binary Numbering System is one in which only 2 symbols are used to display and signify all quantities. The motivation for using the Binary Base-2 Numbering system can easily be seen by observing the various Dual principles that are observed in Nature: Male and Female, Sun and Moon, Light and Dark, True and False, Right and Wrong, Left and Right, Up and down, High and Low, On and Off, 2 Eyes, 2 Ears, 2 Nostrils, 2 Lips, 2 Arms, 2 Hands, 2 legs, 2 Feet, etc.

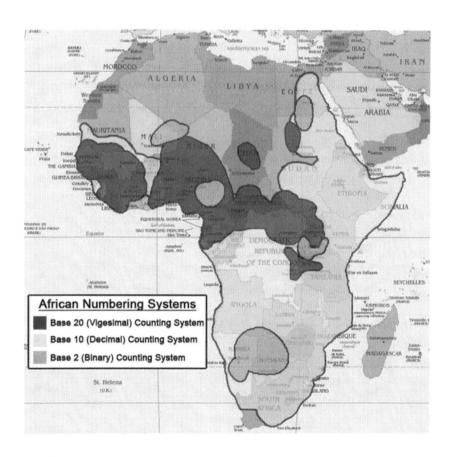

There are pros and cons to the different numbering systems depending on the task being done. The oldest numbering system found in Africa was the Binary base-2 numbering system. Besides the modern uses for the African binary base-2

numbering system in computers, electronics, and digital logic; the African binary base-2 numbering system also had advantages over the decimal base-10 numbering system in terms of counting and performing arithmetic on the fingers of the hands. Using the decimal base-10 numbering system, it is only possible to display quantities as large as 5 using all of the fingers of one hand, and quantities as large as 10 using all of the fingers on both hands. However, using the African binary base-2 numbering system, it is possible to display quantities as large as 31 using the fingers on one hand, and quantities as large as 1023 using the fingers on both hands. This is due to the fact that each finger represents a power of 2.

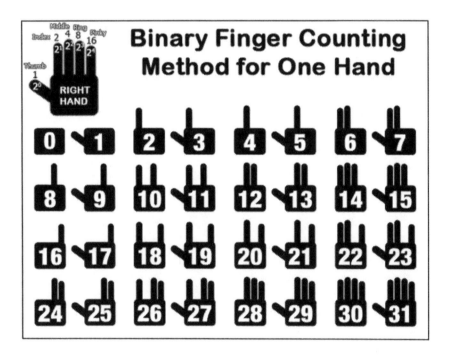

Nine To The Ninth Power of Nine

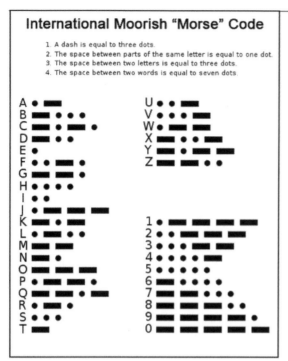

The African binary-base 2 numbering system can also be used to indicate quantities using tone, rhythm, and light. Since the binary base-2 numbering system only needs two indicators to specify a quantity, the basis of music, which is a Beat and a Rest, can be used as a method of conveying Binary data. The "Beat" or the Hit of the drum would indicate 1, on, High, etc and the "Rest" or pause would indicate 0, off, low, etc. This concept of using sound and rhythm to convey binary data was taken to develop "Morse Code". **Morse code** is a rhythmic signal used on the Telegraph to communicate letters and numbers to send messages. The invention of Morse code is attributed to Samuel F. B. Morse; however the concept of using **sound, tone, rhythm, and light** to communicate and convey information over long distance is a practice that has been done in Africa for centuries. In fact, the etymology of the name **"Morse"** is "dark-skinned, **Moorish**" which shows that the concept originated with Black People in Africa.

The applicable concepts of the African Binary base-2 numbering system which were used to develop the Morse (Moorish) code system also served as the inspiration to the development of the **Barcode** and Universal Product Code (UPC) systems which are used to electronically track products, services, retail items, or anything that can be assigned a number.

The first barcodes were constructed by vertically extending the Morse code dashes and dots to form lines of varying width which could then be used to optically encode and transmit binary data using Light.

Above: The words "African Creation Energy" written in Morse code and extended vertically to form a Barcode using African binary base-2 numbering system encoding.

Another modern system that depends on the African Binary Base-2 system is **Braille**. Using the sense of touch, the Braille system enables Blind people the ability to read and write using raised dot characters arranged in any of six positions in two columns for a total of sixty-four (2^6) permutations,

Braille Alpha-Numeric Characters

a	b	c	d	e	f	g	h	i	j
k	l	m	n	o	p	q	r	s	t
u	v	w	x	y	z				
0	1	2	3	4	5	6	7	8	9

The binary base-2 numbering system was also used in Africa to express rational number values as fractions. Six component parts of the famous "Eye of Horus" symbol were used to indicate binary fraction quantities with the entire "Eye of Horus" symbol being an approximation of the quantity "1" that was short by $\frac{1}{64}$. The Left eye of Horus was symbolic of the Moon and associated with the deity **Tehuti**, logic, and thinking while the Right Eye of Horus was symbolic of the Sun and associated with the deity **Ra** and **creativity**. This also corresponds to modern scientific classifications of the Left Side of the Brain being responsible for Logic, and the Right side of the brain being responsible for Creativity.

Supreme Mathematic African Ma'at Magic

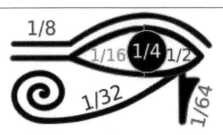

Eye of Horus Egyptian Binary Fractions Analysis:

Exponent Power n	Binary Fraction $1/2^n$	Hekat Value	Symbol	Symbol Description	Sense
1	$\dfrac{1}{2^1}$	$\dfrac{1}{2}$		Nose	Smell
2	$\dfrac{1}{2^2}$	$\dfrac{1}{4}$		Eye Pupil	Sight
3	$\dfrac{1}{2^3}$	$\dfrac{1}{8}$		Eyebrow	Thought
4	$\dfrac{1}{2^4}$	$\dfrac{1}{16}$		Ear	Hearing
5	$\dfrac{1}{2^5}$	$\dfrac{1}{32}$		Tongue	Taste
6	$\dfrac{1}{2^6}$	$\dfrac{1}{64}$		Hand	Touch
SUM TOTAL		= 63/64 = 0.984 ≈ 1		All Information Received via the senses By The Human Mind	

Nine To The Ninth Power of Nine

Each of the six component parts of The Eye of Horus represented a body part that was responsible for delivering sensory information to the Human brain. The binary Eye of Horus fractions were used as a means to measure Volume by ancient Africans. Volume is the amount of **space** that **matter** occupies. In modern terms, Volume is measured in units of cubic meters or liters; however the eye of Horus binary fractions were used to measure volume in units called the **"Hekat"** or **"Heqet"** where the entire eye of Horus was 1 Heqet, and each of the 6 component parts of the eye of Horus represented a binary fraction of the Hekat. Egyptian fractions were also written using decimal base-10 notation where the decimal number symbol was placed under the ⌬ symbol to indicate a fraction. For example, $\overset{\frown}{|||} = \frac{1}{3}$ and $\overset{\frown}{||} = \frac{1}{5}$. The Egyptian decimal fractions were used to measure volume in units called the **"Ro"** or **"Ra"**. African Egyptian mathematicians could convert between units of Heqet or Ra by the conversion factor:

- **1 Heqet = 320 Ra**
- 1/2 Heqet = 160 Ra
- 1/4 Heqet = 80 Ra
- 1/8 Heqet = 40 Ra
- 1/16 Heqet = 20 Ra
- 1/32 Heqet = 10 Ra
- 1/64 Heqet = 5 Ra

Volume (the amount of space occupied by matter) was and is a necessary measurement to make before **Creating** or constructing any physical object. It follows that the names **Heqet** and **Ra** were used for **Creation deities** in Ancient Egypt. Heqet or Hekat, whose name is a feminine version of "heka" or "Magic" (engineering), was a creation goddess associated with "**Birth**" and "**fertility**" and was said to be the wife of another Creator god named "**Khnum**" whose name means "**Create**". **Ra** was the name of the Ancient Egyptian Sun God who had many functions with one of his primary functions being **Creation**. Amongst the ancient African Egyptians **Ra** was said to be the father of **Ma'at**.

African Creation Deity RA

African Creation Deity HEQET

African Creation Deity KHNUM

Nine To The Ninth Power of Nine

In addition to the Binary base-2 and Decimal base-10 Numbering systems; base-5, base-6, and base-8 numbering systems are also used throughout Africa. Another prevalent numbering system found in Africa is the Vigesimal, base-20 numbering system. The Vigesimal base-20 numbering system is one in which only 20 symbols are used to indicate values 0 through 19. In this numbering system, the symbol for 19 would be the highest valued symbol and all other numbers after would be a combination of the 20 symbols. The Vigesimal base-20 numbering system was also used by the Mesoamerican Mayan Native American Indians, as the numerical and mathematical foundation to calculate the **Mayan-Long Count Calendar.**

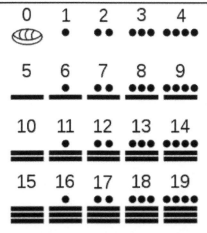

Mayan Vigesimal Base-20 Numbering system, similar to African Vigesimal Base-20 Numbering systems

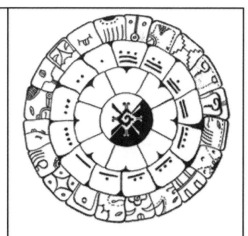

The Mayan Tzolk'in Calendar was calculated using a Vigesimal Base-20 Numbering system that is also used in Africa

The African binary base-2 number system is the number system that is used to mathematically express formulas for Logic which is the systematic method of reasoning. The African binary base-2 number system is the numeric system needed to mathematically determine Ma'at (Truth, Rightness, and Justice), for example with the binary number 0 representing false and binary number 1 representing truth. In terms of mathematical logic and establishing truth (Ma'at) The African Vigesiml base-20 number system enables determination of degrees of truth and degrees of false. For example, in the African vigesimal base-20 number system, if the symbols for quantities 0, 1, 2, 3, 4, 5, 6, 7, 8, and 9 represent "false", then each symbol would indicate a "degree of false". If the symbols for 10, 11, 12, 13, 14, 15, 16, 17, 18, and 19 represent "truth", then each symbol would indicate a "degree of truth". Therefore the African vigesimal base-20 number system can be used to mathematically determine Statistics and Probability or that which is probably true or probably false.

4. THE MATH OF MA'AT

The African deity Ma'at represented the concepts of Order, Balance, Truth, and Rightness. The principles that Ma'at represents have a direct relationship with Mathematics. The concept of "Order" deals with rank, priority, sequence, arrangement, symmetry, harmony, position, and importance. Order is established mathematically by way of **"Ordinal Numbers"** using Mathematical **Set Theory**. Ordinal Numbers are simply numbers used for establishing rank, order, sequence, importance, or position. Ordinal Numbers are usually written with an ordinal indicator sign, for example the **numero** sign (№), to denote that the number is an Ordinal Number. Set Theory or Group Theory is the branch of Mathematics that studies collections of Objects called "Sets" or "Groups".

It is through the study of Ma'at (Order) through Set Theory that it was mathematically proven that there are different types of "Infinities" and then an "Infinity of Infinities" which would be considered **"Actual Infinity"**. The mathematical concept of the **"Infinity of Infinities"** is also referred to as the **"Galaxy of Galaxies"** and the **"Universe of Universes"**. The ancient Africans simply called this concept **"The All"**, and it was said symbolically that "Ma'at **set** the **Order** of the **Universe** and the **Cosmos**".

The study of the concepts of Infinity, the All, and/or the Universe is called "**Cosmology**". The word "Cosmology" comes from the words "**Cosmos**" meaning Ma'at, "order" or "universe" and the suffix "-ology" meaning "the study of". The concepts of Cosmology (the study of Ma'at, Universal Order) form the foundation of all Cultures, religions, philosophies, and societies. While the details of the Cosmology story may differ from culture to culture, amongst African cultures, general Cosmology themes are consistent.

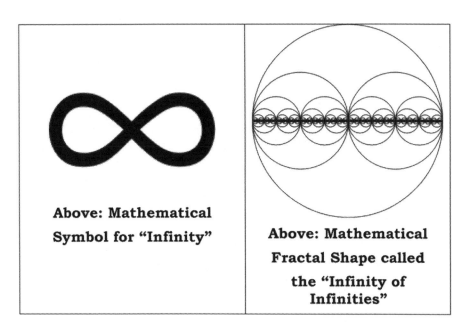

Above: Mathematical Symbol for "Infinity"

Above: Mathematical Fractal Shape called the "Infinity of Infinities"

African Cosmology stories usually pertain to **Creation Stories** that describe how the Universe and everything were created into existing order from Chaos, and how the universe will eventually be destroyed (Eschatology) and return to Chaos from Order, and continuously repeat this cycle of Order-to-Chaos and Chaos-to-Order. The principles of Cosmology Creation and destruction stories are depicted geometrically with figures called Cosmograms. Because of the importance of the Cosmology principles that Cosmograms represent, the geometry of Cosmograms are usually considered **"Sacred Geometry"** amongst their respective cultures.

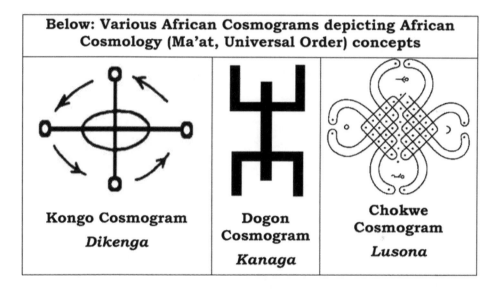

Below: Various African Cosmograms depicting African Cosmology (Ma'at, Universal Order) concepts

Kongo Cosmogram *Dikenga*

Dogon Cosmogram *Kanaga*

Chokwe Cosmogram *Lusona*

The word **"Cosmos"** meaning "Order" or **"Ma'at"** can also be found in the word **"Cosmetology"**. The word cosmetology literally means "to arrange or put in order" (to perform Ma'at) but is used to refer to the science and art of beautification which includes hairstyling, skin care, cosmetics, manicures, pedicures, etc. Cosmetology, applying Ma'at, as the science and art of beautification is an art that African women have been doing for thousands of years. The mathematical process of applying Ma'at, Order and Arrangement for beautification (called Cosmetology in modern language) involves creating symmetric Geometric patterns that are aesthetically pleasing, proportional and balanced. The mathematical principles of applied Ma'at for beatification (Cosmetology) are most evident in the geometric patterns found in African hair styles. It has been scientifically shown that geometric facial symmetry is a trait that humans find most attractive and desirable; therefore one of the goals of applying Ma'at in Cosmetology is to make the face appear symmetrical. Ma'at in the form of Cosmetology leads to beautification, attraction, and pro-creation, which further illustrates the association between Ma'at and Creation. In sounds and tones, the application of Ma'at manifests as musical harmony. In building and construction, the application of Ma'at is manifested as solidity and stability. The understanding and Application of the principle of Ma'at is essential in any type of Creation.

Above: Various mathematic geometric patterns found in Applied Ma'at techniques called "Cosmetology"

The Ma'at principle of "Balance" is another concept that is determined mathematically. Balance or Equality is established by Measurement or quantitative comparison. Measurement is nothing more than the comparative act of counting and quantifying one thing relative to, or in terms of another thing. For example, weight or Mass is measured by counting how many of one type of thing balances with another type of thing. Volume is measured by counting how many of one type of thing fits inside of another type of thing. Time and Existence is measured by counting how many similar events can take place relative to another event. Therefore, Ma'at as Balance is determined by the Mathematical Arithmetic process of measurement through counting. In order to measure, you must have at least 2 things: an object to measure and another object to serve as the reference to quantify the object being measured.

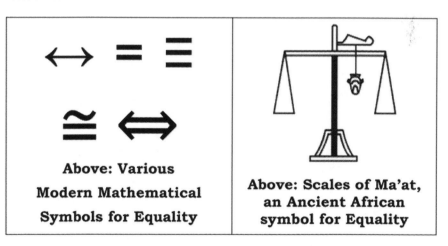

Above: Various Modern Mathematical Symbols for Equality

Above: Scales of Ma'at, an Ancient African symbol for Equality

The mathematical concept of measuring to determine balance was depicted by the ancient Africans symbolically with the "Scales of Ma'at." A scale is used specifically to measure weight. Weight is defined scientifically as the magnitude of the Force that must be applied to an object to obtain equilibrium in a gravitational field. The concept of "weight" can be expanded to generally refer to "the quantitative Magnitude of any force that is used to obtain Equilibrium, Equality, or Balance". Therefore, the "Scales" as a symbol of determining Ma'at have a broader relevance beyond measuring the "weight" of an object, but can be used to symbolically refer to any force that when applied as a solution to a problem; equalizes or balances the equation. The scales are said to be "Balanced" when the scales are stable, and the two objects being measured are "Equal" or in equilibrium. Mathematically, the concept of "Balance" is written using **equations**. An equation is a Mathematical statement written in symbols that shows that two things are **Equal**. The two things being "measured" or compared in a mathematical Equation are separated by an "Equals" sign just as the two objects being measured or compared by Ma'at are separated on opposite sides of the symbolic scales. To solve a mathematical equation is modern terminology for the ancient African concept of determining Ma'at. Determining Ma'at (solving equations) is the process of finding any and all solutions, values, quantities, functions, or variables that satisfy the condition of Equality.

Solving Equations = Determining Ma'at

There are a number of different qualities, attributes, and properties for which the Equality of two things can be determined. If two things are quantitatively equal for all qualities, attributes, and properties, then the two things are said to be Identical or the same. This is because an Equality relation is an Identity relation – that which readily identifies and defines something. The ability to accurately, correctly, rightly, and truthfully define something by way of equivalence and equality goes into perhaps the most important principle that Ma'at represents which is Truth.

Ma'at as the principle of truth (accuracy, correctness, rightness, righteousness, fact, reality, etc.) can also be determined mathematically. The mathematical process of determining Truth is called **Logic**. Logic is the step-by-step process of **Reasoning**, and is also called the study of reasoning. There are only two Logic values (called **Truthbearers**): True and False. For this reason, the Mathematical study of logic only requires two symbols, and therefore, the binary number system is used in mathematic Logic equations.

The mathematics of Logic is the fundamental and essential driving force of modern computers, computer programming, and programmable automation electronic devices. Modern mathematicians trace the study of Logic back to the Greek philosopher named **Aristotle**. Aristotle and other Greek philosophers called Logic by its etymological origin **Logos**. Greek philosophers viewed Logos (or Logic) as the "spoken word that was the source of the order of the cosmos". However, it is known that Aristotle stole his ideas and concepts from the Ancient Egyptians. This concept of the Logos being the creative utterance that brought the order of the cosmos into existence was already known to be in existence in Africa for hundreds of years prior to Aristotle and the other Greek philosophers.

Above: African Creation Deity Ptah (Logic and Reason) depicted standing on a 4-sided platform which represented Ma'at (Truth and Order).

It is written in the **Memphite Theology** which dates back to 4000 B.C. and predates the Greeks and was also carved on the "**Shabaka Stone**" during the **Nubian 25th dynasty of Egypt**, that the African deity **Ptah** was "*The Opener who through creative utterance, called the universe into existence after having imagined creating in his heart and speaking it with his tongue*". By definition of function, the role of the African deity Ptah in African philosophy is equivalent to the **Logos** (creative utterance or **Logic**) and the **Nous** (intention of the Mind or **Reason**) in Greek Philosophy. Depictions of the African deity Ptah show him standing on a 4-sided platform foundation which represents Ma'at. Mathematically, this depiction symbolizes the importance and relationship of Logic (Ptah) to Truth (Ma'at), as if to say "*Truth is the foundation of Logic and Reason*".

The Mathematical process of symbolically writing the formulas and forms of logic is studied in the fields called **Propositional Calculus, Mathematical Logic,** and **Boolean algebra**. Propositional calculus is a modern mathematical system of translating the Logic of propositional statements, concepts, and arguments into equations and formulas that can be connected and combined using logical operators. Each logical operator is called a "Truth Function" (Ma'at function) and is commonly expressed as a binary operation on two values. Boolean algebra extends the Logical operations into an

Algebraic system of truth values that has applications in Mathematics, Philosophy, computer science and electronics.

It is said that Boolean algebra was invented by a man named "George Boole", however, it is interesting that the word "Boole" is related to the word "Boule" which comes from a Greek word meaning "to decide or deliberate", and "Boolean Algebra" is a system of Mathematically "deciding or deliberating" what is true via logic. Also, the word "Boule" refers to a Crystal of silicon which is used as the material to construct semi-conductors which serve as the electrical Logic-components on which Boolean algebra is performed via programming.

The Logic operations of Boolean algebra are performed on the two symbols in the Binary Number system, 0 and 1, with 0 representing "false" and 1 representing "true". The three basic logic operations in Boolean algebra are:

- Logical **AND** (also called "Conjunction" and logic multiplication)

- Logical **OR** (also called "disjunction" and logic addition)

- Logical **NOT** (also called "complement," "inversion," and logic negation)

Nine To The Ninth Power of Nine

The three basic logic operations can be combined mathematically to form other logic operations including: Implication, Contradiction, Equivalence, "Not both (NAND)," "Neither...nor (NOR)," denial, tautology, and others. The root of the word "Logic" is "Logos" and the term "Logos" or "Logo" is also used to refer to a graphic, picture, or symbol used to represent or signify some idea or concept. In Boolean algebra, the Logos or symbols of the Logic operators are expressed using mathematical notation, truth-tables, Venn diagrams, and Logic gates. "Logic gates" are a specialized Logos of logic used in electronic schematics and diagrams to indicate Logic operations. Venn diagrams are logos of logic that are not only used in Logic and computer science, but also in statistics and probability.

Math Logos of Logical Operators

	Logical AND	Logical OR	Logical NOT
Math Notation	$A \cap B$ $A \wedge B$ $A * B$ A AND B	$A \cup B$ $A \vee B$ $A + B$ A OR B	$\bar{A}$ $\neg A$ $-A$ NOT A
Truth Table	AND: A=0,B=0→0; A=0,B=1→0; A=1,B=0→0; A=1,B=1→1	OR: A=0,B=0→0; A=0,B=1→1; A=1,B=0→1; A=1,B=1→1	A=0 → NOT A=1; A=1 → NOT A=0
Venn Diagram	(intersection shaded)	(union shaded)	(complement shaded)
Logic Gate	(AND gate)	(OR gate)	(NOT gate)

The logical operators AND, OR, and NOT are equivalent by definition of function to several symbols in the **Nsibidi** script developed by the **Ekpe** secret society amongst the **Efik** people of **Nigeria**. Adjectives used to describe the Ekpe secret society are "Clerks," "Recorders," and "Scribes." The word "Ekpe" means "**Leopard.**" This secret society of **Scribes** called the "**Leopards**" (which shows a clear association with the Egyptian deity **Seshat**) developed a script called **Nsibidi** which contains several Logos or symbols that could be considered African equivalents to the various Logos of the Logic operators AND, OR, and NOT.

African Logos of Logical Operators

BELOW: Nsibidi symbols of the Ekpe (Leopard) African Scribe secret society. These symbols are equivalent to various modern Logic logos by definition.

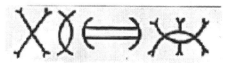

Union or Marriage

Logical AND

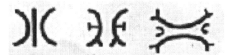

Separation or Divorce

Logical OR

Mirror or Reflection

Logical NOT

Supreme Mathematic African Ma'at Magic

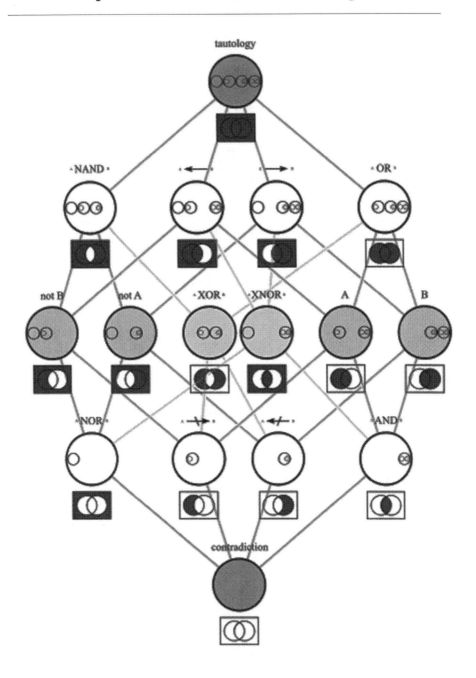

Above: A Boule (Crystal Lattice) diagram showing the various Venn Diagram combinations of Logic Operators in Boolean Algebra – the Mathematics of Logic and Reason

In computer programming, Logic operators are combined with **conditional statements** such as **"If...Then...Else"** to determine which action should be performed according to the Logic programming. The "42 Laws of Ma'at" also known as the "42 Negative Confessions of Ma'at" are a set of 42 conditional statements prefaced with the Negation logical operator that were said to be asked of the deceased person in the Halls of Ma'at when the heart of the deceased person was weighed against the feather of Ma'at. If truth was determined for all 42 Negative confessions of Ma'at, then the deceased person was permitted to leave the Egyptian underworld called the Duat. The interesting thing about the 42 Negative Confessions is that they are written in the same way a computer program would be written to perform the same task. That is, the propositions in the 42 Confessions use the Negation (usually expressed by "Not") operation in Mathematical Logic (Ma'at magical Reasoning). The Truth Function (the Feather of Ma'at) is used to determine if the proposition is True or False.

Just as the principles of Ma'at can be expressed mathematically, called ***"the Math of Ma'at"***, Mathematical operations also follow a Ma'at or Order called ***"the Ma'at of Math"***. The Ma'at of Math would be the Order which Mathematical operations are performed in an equation. In computer programming, it is essential to understand the Ma'at of Math (the order of mathematical operations) to understand <u>what</u> Mathematical operation occurs <u>when</u>. Given the mathematical operations of Exponentiation, Multiplication (including Division), and Addition, (including subtraction), the Ma'at or Order of Mathematical operations is:

1. Exponentiation (and roots)
2. Multiplication (and division)
3. Addition (and subtraction)

42 LAWS OF MA'AT
(42 Negative Confessions of Ma'at)

1	I have not committed sin.
2	I have not committed robbery with violence.
3	I have not stolen.
4	I have not slain men or women
5	I have not stolen food.
6	I have not swindled offerings.
7	I have not stolen from God/Goddess.
8	I have not told lies.
9	I have not carried away food.
10	I have not cursed.
11	I have not closed my ears to truth
12	I have not committed adultery.
13	I have not made anyone cry.
14	I have not felt sorrow without reason
15	I have not assaulted anyone
16	I have not been deceitful.
17	I have not stolen anyone's land
18	I have not been an eavesdropper
19	I have not falsely accused anyone.
20	I have not been angry without reason.

42 LAWS OF MA'AT
(42 Negative Confessions of Ma'at continued)

21	I have not seduced anyone's wife.
22	I have not polluted myself.
23	I have not terrorized anyone.
24	I have not disobeyed the Law.
25	I have not been exclusively angry.
26	I have not cursed God/Goddess.
27	I have not behaved with violence.
28	I have not caused disruption of peace.
29	I have not acted hastily or without thought.
30	I have not overstepped my boundaries of concern.
31	I have not exaggerated my words when speaking.
32	I have not worked evil.
33	I have not used evil thoughts, words or deeds.
34	I have not polluted the water
35	I have not spoken angrily or arrogantly.
36	I have not cursed a God.
37	I have not placed myself on a Pedestal.
38	I have not stolen what belongs to God/Goddess.
39	I have not stolen from or disrespected the deceased.
40	I have not taken food from a child.
41	I have not acted with insolence.
42	I have not destroyed property belonging to God/Goddess.

5. THE MATH OF MYTH

The word "**Myth**" comes from the Greek word "Mythos" meaning "Thought". The word "**Math**" also comes from a Greek word meaning "Thought". However, what separates Myth from Math is the fact that Math concepts are a collection of thoughts that have been verified and proven as factual, actual, and true, whereas Myth concepts are a collection of thoughts that may or may not be true. Therefore, Myths and Mythology concepts, ideas, and stories always contain a degree of **uncertainty**. Myths are accepted and believed by people because even though they may not be 100% verified and proven to be accurate and true, the explanations that Myths offer are **likely** to be true and **probably** accurate given the information that people have at the time.

The concepts that form the foundation of Myth - uncertainty, likelihood, probability, doubt, assumption, speculation, approximation, guessing, percentage, prediction, randomness, and chaos – can all be approached mathematically through Probability and Statistics. Probability and Statistics as the "Math of Myth" comes into play when the mental reasoning process goes from deduction (making specific conclusions from specific evidence) to induction (making general conclusions from specific evidence). When performing inductive reasoning to make general broad statements based on certain specific forms of evidence and observations, statistics and probability (math) is used to say (myth) "since the general conclusion that is being drawn from this specific evidence is right <u>most</u> of the time, then the general conclusion is <u>probably</u> right for all of the evidence all of the time." Another term for the "Math of Myth" is "Approximation".

Approximation is the **inexact** representation of something that is still close enough to be useful. **Approximation** is a lesser form of **Equality**. Approximations are close, ambiguous, comparable, alike, unclear and inexact, whereas Equality is precise, exact, same, and clear. The mathematical symbol for approximation is ≈ or ≋. This mathematical symbol for approximation resembles the ancient Egyptian hieroglyphic for water〰, which is pronounced **Nu** or **Nun**. In ancient Egypt, **Nun** and his female counterpart **Nunet** represented the concepts of "the primordial abyss" and "the chaotic waters" and were metaphors for the concept of "Chaos" that preceded creation Order (Ma'at).

Above: Various Mathematical Symbols for Approximation

Above: Various Hieroglyphic for Nun, the Ancient African symbol for Chaotic Waters

Mathematically, chaos is described as a state that lacks Predictability and therefore Approximations and probability are necessary. The interesting thing about the mathematical study of Chaos, probability, and statistics is that even though these concepts are considered random and without order or predictability, patterns can be found which could be considered the "Order of Chaos".

Nun

Nunet

Above: Nun and Nunet, Ancient African deities symbolizing "Primordial Abyss" and "Chaotic Waters" – metaphors for Chaos

The "Order of Chaos" or the "Pattern of Randomness" in existence has been mathematically studied in a field called **Chaos Theory**. Chaos theory is a deterministic and ordered mathematical model or equation that is extremely sensitive to change and thus can have very different outputs (effects) for slightly different inputs (causes). The wide variability in the outputs that Chaos theory models generate for slightly different inputs makes predicting the outcome of the model difficult, if not impossible, and thus the result is declared Random or Chaos even though the method which produced the result has a determined and ordered form and formula. The same formula concepts in Chaos theory can be applied in computer programming to perform "Random Number Generating", for example to generate a "random" ID or password. Chaos theory is basically a modern Philosophical, Scientific, and Mathematic reframing of the Ancient African ideas that everything is related, and every cause has an effect.

Chaos theory is also related to the statistical concepts of **Game Theory** and **Decision Theory** which mathematically study the probability and statistics of the causes and effects of **Choice** and "**Free Will**". The idea that a formula, equation, computer program, card deck, coin toss, dice throw, or other form of "Randomization" agent could produce a result that is somehow chaotic, random, or unpredictable is antithetical because all of the aforementioned randomization agents obey some form of law and order. Even if accurate prediction is difficult due to the many variables that may contribute to the outcome, accurate prediction is not impossible if all of the information is taken into consideration. In reality, the terms "Order" and

"Chaos" are two sides of the same thing that are subjectively distinguished by human mental comprehension capabilities.

The relationship between Order and Chaos can be seen using Mathematics to geometrically graph a result of Chaos theory called the "**Butterfly Effect**". "*The Butterfly Effect*" is a concept used in Chaos theory to denote how the flap of a butterfly's wing on one side of the world could cause a chain of events that eventually lead to a tornado on the other side of the world, but more specifically relates to a geometric graph that looks like the wings of a butterfly and depicts the variability and cyclic nature of outcomes for various inputs. The butterfly effect fractal geometric shape from Chaos theory has a similar figure-eight (∞) geometric form as the "Infinity-of-Infinities" fractal shape from the study of Order in Set Theory.

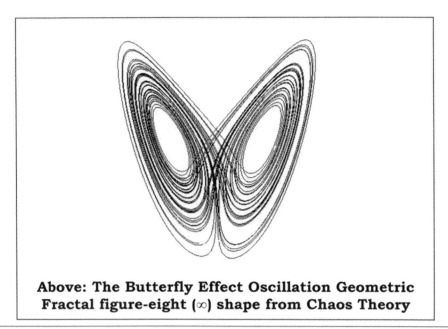

Above: The Butterfly Effect Oscillation Geometric Fractal figure-eight (∞) shape from Chaos Theory

The principles of "Chaos" are also mathematically applied to produce and generate various Fractal shapes found in Nature. A "**Fractal**" is a rough geometric shape that when divided and separated into parts, each part is just a smaller copy of the larger shape. Fractal geometric shapes exhibit the Mathematical property called "Self-Similarity" in which a smaller component part of a larger object is exactly or approximately similar to the larger object. Fractal shapes are not only prevalent in nature, but are also frequently found in the village layout and Architectural design in Africa, as well as in African sculptures, textiles, and art. It has been said that the word "Africa" comes from the word "Fraction" which is also related to the word "Fractal" and "fracture" meaning "to break apart" or divide. Considering the definition of a Fractal as being something where the component parts are replicas of the whole, and vice-versa, mathematically, it would make sense for the word Africa to be related to the word Fractal as well as Fractal shapes to be prevalent in Nature and African culture because the Fractal geometric shape is the Mathematical and geometric application of the ancient African concept of: "**As Above, So Below**". What is interesting is that the Geometric Mathematical concepts that are used by Nature and by African people to Create structure and form is classified under the modern Mathematical concept of "Chaos" even though these forms are Ordered by Nature.

Supreme Mathematic African Ma'at Magic

Fractal shapes found in African Kente Cloth said to have been inspired by the weaving of a Spider's web

Natural Fractal shape found in the design of a spider's web

Natural Fractal shape formed from the Fracture or Breaking of Glass

The need and ability for humans to foretell and predict the future, as well as retell and retrace the past has been a function that has existed in societies throughout the ages. Various titles such as fortune tellers, soothsayers, seers, prophets, oracles, forecasters, and analyst have been given to people whose responsibility it is to foretell the future and retrace the past. The mental process that these individuals have used to be able to provide this information is "inductive reasoning" and the Mathematic process of inductive reasoning is probability and statistics. Probability and statistics is the mathematical process that enables the Weather forecasters of today to predict future weather patterns. Probability and statistics is also the mathematical process that enables archeologist of today to retrace the past and provide dates to their archeological finds. Mathematical models using existing data and information are developed that enable prediction and retracing. In ancient Africa, the "foretelling" and "retelling"

Mathematical models were the various cosmograms of universal cycles and their associated cosmology. In modern times, the "foretelling" and "retelling" Mathematical models are found in probability and statistics.

Venn Diagrams are one form of Mathematical model that is used in probability and statistics. Venn Diagrams are used in Logic as well as in Probability, however, when used in logic, the numerical values found in Venn Diagrams are only 1 or 0, but when used in Probability when "Truth" is not certain, the numerical values found in Venn diagrams are Fractions of 1. Venn diagrams are said to have been developed by a man named John Venn, however, another interesting coincidence is that the name "Venn" comes from the word "Win", meaning to <u>struggle</u> for success, and Venn Diagrams are used as mathematical decision models to determine success.

Probability and statistics as the "Math of Myth" are mathematical processes that are performed without having complete information, all the data, or all the facts. It is almost impossible for any person to obtain <u>all</u> the information on a subject, so inductive mental processes or myths are natural when coming to conclusions and arriving at explanations. Probability and Statistics as the "Math of Myth" enables the individual to come to the most accurate conclusion, explanation, and/or prediction for the evidence and information that they have obtained, even if the information that has been obtained is limited. But, for the purpose of completeness, accuracy, rightness, exactness, validity, and reality, if probability and statistics are used via inductive reasoning to come to conclusions, then it is necessary that the conclusions, and methods that were used to arrive at the conclusions, be revisited and revised whenever more information and evidence is obtainable.

Myth stories, allegories, and parables are usually metaphors for which certain principles and ideas in the specific story can be transferred using Inductive Reasoning to other more general principles and ideas which can be applicable to other scenarios, circumstances, situations, experiences, and examples outside of the story. The ability to use inductive reasoning to go from a <u>specific</u> observation, evidence, and/or experience to a more <u>general</u> conclusion about the past, present, or future or to provide esoteric and intuitive explanations is called **Insight**, and the mathematical processes of Insightful Inductive Reasoning is Statistics and Probability.

6. THE MYTH OF MATH: NUMEROLOGY

Numerology is the mental process of attempting to find a relationship between numbers and reality. The hypothesis, beliefs, and ideas that are formed in Numerology are by definition the beginning stages of any Mathematical concept, and what brings a concept from the realm of Numerology to the Real of Mathematics is the ability of the concept to pass **rigorous deductive reasoning**.

The esoteric hypothesis, beliefs, speculations, and ideas in Numerology have been expanded from just numbers to assigning interpretations and explanations to Mathematical operations such as Arithmetic, Algebraic Equations, Numerical permutations, groupings, and orders, as well as assigning Numerical values to letters to obtain a Numerological determination of words.

Almost every person is a Numerologist to some degree in that they assign a significant meaning to numbers, whether that number is a "Lucky Number" (7 for example), "Unlucky Number" (6 or 13 for example), Date (Birthday for example), Scientific or Mathematical Constant (Pi = 3.14159265 or Phi = 1.61803399 for example), Fibonacci sequence, Temperature, pH value, etc. Even organized religions that once denounced Numerology have began "Numerological divination practices" such as the Biblical scholars' **"Bible Code"** or Islamic Scholars searching for significance of the **Number 19** in **"Quran Code"**. Bibliomancy is the use of books in divination practices and is commonplace in almost every religion worldwide.

Because of the fact that most people have some degree of a Numerological tendency when it comes to numbers, Numerology is addressed here to show the African origin of Numerology and also to motivate the **transformation** of **Potential energy** spent in Pure Speculative Numerology into Active **Kinetic Energy** engaged in reasonable, operative, and applicable Mathematics.

Numerology (speculation) gives birth to Mathematics (operation) through reason. Numerology is the beginning stages of Mathematic thought processes which eventually become Mathematic Activities. Numerological ideas and concepts can also take place after Mathematic concepts are developed through unreasonable speculation, assumptions, and conjecture.

Nine To The Ninth Power of Nine

	Common Numerological Qualitative, Subjective, and Speculative Associations for Numbers
0	Void, non-being, Naught, the originator and container of All
1	Life/Death, Identity, Aught, Primordial Being, Wholly One, Form, Initiating, Unity, Creation, Oneness, Individual, Aggressor, Yang, sure
2	Container, Formless Form, Polarity Division, Two to Tango, Content, Cooperation, Duality, Dividing, Balance, Union, Receptive, Yin, easy
3	Movement, Synthesis, Harmony, Agreement, Expression, Relationship, Power, Divine Perfection, Communication, interaction, Neutrality, live, number of the Leviathan
4	Resistance, Solidity, Material Order, Mother Substance, Distinction, Foundation, Solidity, Creation, unlucky in Cantonese since the pronunciation of 4 is a homonym with the word for death or suffering
5	Life, Love, Regeneration, Identity, Expansion, Growth, Sensuality, Grace, Action, Restlessness, the self, me, myself
6	Sex, Union, Perfection, Created World, Structure Function, Value, Nurturing, Movement, Perfection, Carnal man, Reaction, flux, Responsibility, easy and smooth all the way
7	Indeterminate Probability, Critical Time Chance, Wisdom, Enchanting Virgin, Consequence, Understanding, Attainment, Mysticism, Spiritual Perfection, Thought, consciousness, a vulgar word in Cantonese
8	Sum of Possibilities, Harmonic Sum, Auspicious, Periodic Renewal, Invariance, Practical, Justice, New Birth, Power, sacrifice, sudden fortune, prosperity
9	Cell, Structure, Limit End, Magnified Three, Horizon, Variance, Humanitarian, Achievement, Finality, Highest level of change, long in time
10	Existence, Perfection Wholeness, Completion, Beyond, God's Law, Rebirth

Just as Mathematics has its origins in Africa, Numerology also originated in Africa. Ancient Africans used specialized process (rituals) to gain insight, understand Nature, and solve life's problems using Numbers and Mathematics. Attempting to replicate a Mathematical process (ritual) for religious purposes without knowing the Mathematical purpose or reason gives way to what is called **Numerological Divination**. Numerological Divination attempts to gain insight from God into a question or situation by using numbers. It is clear to see how Numerology is Mathematics that is absent of reason. Many African Mathematic processes were observed and copied by the early Greek Mathematicians/Numerologist, but the reason was not fully comprehended, and thus Numerology (Mathematics without reason) spread throughout the world. There are many Numerology Systems due to the many speculations and interpretations of the real meaning of Ancient African Mathematics, Numbers, and symbols. Most Modern Numerological systems can be traced back to **Pythagoras** and other ancient Greek Numerologist who studied **Egyptian Mathematics**. Other origins of Numerological systems are the **Kabbalah Hermetic numerology** that can be traced back to the African deity **Tehuti** and Asian **I-Ching/Taoism numerology** that can be traced back to **African Ifá system of the Yoruba people**.

In addition to the mystical and esoteric explanations that Numerologist place on Numbers and other Mathematical

Objects, Numerologist also assign Number values to letters of the Alphabet in an attempt to find a deeper meaning for words, names, and letters. This practice of adding up the number values of the letters in a word to form a single number for the purpose of Numerological speculation and interpretation is called **Isopsephy** when performed with the Greek and Latin alphabet, and it is called **Gematria** when performed with Hebrew and Semitic languages. Gematria Numerological practices are what motivated and inspired mathematical analysis of the Bible called "The Bible Code" and mathematical analysis of the number 19 in the Quran. **"The Bible code"** (also known as the Torah code) is a series of mathematical algorithms and number combinations that when performed on the Christian religious book called the Bible supposedly demonstrates words or phrases of a prophecy from God. The number 19 code in the Quran is a series of Mathematical algorithms based on the Number 19 that have supposedly been placed in the Islamic religious book called the Quran by God. All of these religious numerology practices fall under the definition of Numerological Divination.

| Arithmancy ||||||||||
| (Numerological Isopsephy Associated with Greek Numerologist/Mathematician Pythagoras) ||||||||||
1	2	3	4	5	6	7	8	9
A	B	C	D	E	F	G	H	I
J	K	L	M	N	O	P	Q	R
S	T	U	V	W	X	Y	Z	

Supreme Mathematic African Ma'at Magic

Gematria Numerology		
Hebrew	**Glyph**	**Number Value**
Aleph	א	1
Bet	ב	2
Gimel	ג	3
Dalet	ד	4
He	ה	5
vav	ו	6
Zayin	ז	7
Het	ח	8
Tet	ט	9
Yod	י	10
Kaph	כ	20
Lamed	ל	30
Mem	מ	40
Nun	נ	50
Samech	ס	60
Ayin	ע	70
Pe	פ	80
Tsadi	צ	90
Kuf	ק	100
Resh	ר	200
Shin	ש	300
Tav	ת	400
Kaph Sofit	ך	500
Mem Sofit	ם	600
Nun Sofit	ן	700
Fe Sofit	ף	800
Tsadi Sofit	ץ	900

Nine To The Ninth Power of Nine

Isopsephy Numerology			
Greek Letter	**Name**	**Transliteration**	**Number Value**
A α	Alpha	A	1
B β	Beta	B	2
Γ γ	Gamma	G	3
Δ δ	Delta	D	4
E ε	Epsilon	E	5
F ϝ	Digamma (Stigma)	W	6
Z ζ	Zeta	Z	7
H η	Eta	E	8
Θ θ	Theta	Th	9
I ι	Iota	I	10
K κ	Kappa	C	20
Λ λ	Lambda	L	30
M μ	Mu	M	40
N ν	Nu	N	50
Ξ ξ	Xi	X	60
O o	Omicron	O	70
Π π	Pi	P	80
Ϙ ϙ	Qoppa	Q/H	90
P ρ	Rho	R	100
Σ σ	Sigma	S	200
T τ	Tau	T	300
Υ υ	Upsilon	Y	400
Φ φ	Phi	Ph	500
X χ	Chi	Ch	600
Ψ ψ	Psi	Ps	700
Ω ω	Omega	O	800
ϡ ϡ	Sampi	Ss	900

The speculative science of Astrology has also become linked with Numerology by associating the characteristics of celestial bodies and Zodiac constellations to number values.

	Zodiac Numerology	
1	Aries - (Cardinal Fire)	♈
2	Taurus - (Fixed Earth)	♉
3	Gemini - (Mutable Air)	♊
4	Cancer - (Cardinal Water)	♋
5	Leo - (Fixed Fire)	♌
6	Virgo - (Mutable Earth)	♍
7	Libra - (Cardinal Air)	♎
8	Scorpio - (Fixed Water)	♏
9	Sagittarius - (Mutable Fire)	♐
10	Capricorn - (Cardinal Earth)	♑
11	Aquarius - (Fixed Air)	♒
12	Pisces - (Mutable Water)	♓

	Astrology Numerology	
0	**Pluto** - transformation and regeneration, depth and intensity	♇
1	**Sun** - conjunction, egocentric, loner	☉
2	**Moon** - opposition, cooperation, emotion, feeling, relationship	☽
3	**Jupiter** - trine, educated, wise, happy, cash flow	♃
4	**Uranus** - eccentric, unusual, inventive, ingenuity	♅
5	**Mercury** - quincunx, communicative, witty, deceptive	☿
6	**Venus** - sextile, pleasant, harmonious, artistic, Tactful, and diplomacy	♀
7	**Neptune** - spiritual, mystical, visionary, seer, idealism, surreal, unreal	♆
8	**Saturn** - solid, stable, lessons learned	♄
9	**Mars** - the square, forceful, dominating, controlling, stability, building, creating	♂

Supreme Mathematic African Ma'at Magic

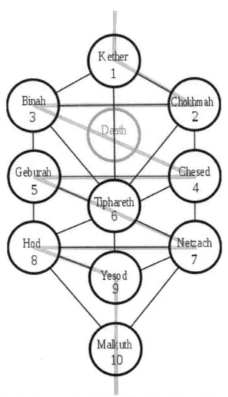

	Kabbalah Numerology
1	Keter (will or crown)
2	Chochmah (wisdom)
3	Binah (understanding)
4	Chesed, Gedolah, or Gedulah (mercy or loving kindness)
5	Gevurah, Din, Pachad (justice, fear, severity or strength)
6	Tiferet (harmony or beauty)
7	Netzach (victory)
8	Hod (glory or splendour)
9	Yesod (power or foundation)
10	Malkuth (kingdom)

Nine To The Ninth Power of Nine

In addition to the Mathematics and Science that the Pythagoreans associated with Numbers and Numerical Operations, they also attributed esoteric and mystical meanings to Numbers through a system called "Pythagorean Numerology". The numerological and

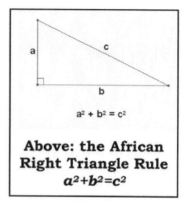

Above: the African Right Triangle Rule $a^2+b^2=c^2$

philosophical doctrines as well as the mathematic concepts and ideas such as the **"Right Triangle Rule"** that have become associated with Pythagoras were in fact presented to Europe by Pythagoras after he learned about and studied the concepts in the African Egyptian cities Heliopolis, Thebes, and Memphis. Similarities between Pythagorean Numerology and Egyptian Numerology are easily identifiable when the two doctrines are compared and contrasted. The application of the mathematic "Right Triangle Rule" can be seen in the construction of the many Pyramids throughout Egypt and Nubia that pre-date Pythagoras. In Numerology, another triangle was significant to Pythagoras and his followers called the **"Tetractys"**. The Tetractys was a triangle composed of 10 points in four rows that symbolized the four elements Earth, Air, Fire, and Water. The concepts of the Tetractys also originated in Egypt where the first row represented a Point, the second row represented a Line, the third row represented a Plane, and the fourth row represented a 3-dimensional Tetrahedron defined by four points which had esoteric significance in the science of creation.

Supreme Mathematic African Ma'at Magic

Pythagorean Numerology Table

#	Name	Description	Symbol
1	Monad	The Point, Oneness, God, the first, totality, the All, whole, the universe, noble number, the sum of the parts considered as a unit, mind, hermaphrodism, odd and even, the receptacle of matter - because it produces the duad (material), chaos, obscurity, the origin of all the thoughts in the universe	(circle with central dot)
2	Dyad	The Line, Twoness, otherness, matter, demiurge, divine mind (nous), division, opposition, genius, evil, darkness, inequality, instability, movability, boldness, fortitude, contention, matter, dissimilarity, Audacity, opinion, fallacy, diffidence, impulse, death, motion, generation, mutation, division, marriage, soul, science, contradistinction, illusion,	(two overlapping circles with line)
3	Triad	Surface, beauty, equilibrium of unities, friendship, peace, justice, prudence, piety, temperance, virtue, wisdom, understanding, knowledge, music, geometry, and astronomy, terrestrials	(triangle in vesica)
4	Tetrad	Solid, Fire, justice, as the primogenital number, the root of all things, the fountain of Nature and the most perfect number, God is expressed as a tetrad, the Number of Numbers, the soul of man consists of a tetrad, the four powers of the soul being mind, science, opinion, and sense	(square/diamond in circles)
5	Pentad	Pyramid, Water, a pentagram symbol was used by the Pythagoreans as a secret sign to recognize each other, life, power and invulnerability, symbol of light, health, and vitality equilibrium, because it divides the perfect number 10 into two equal parts	(pentagram in circles)
6	Hexad	Double triangle, Earth, Infinite divisibility, perfection of all the parts, the form of forms, the articulation of the universe, the maker of the soul, marriage because it is formed by the union of two triangles, time and duration, health is equilibrium, a balance number, the world, hexagon	(hexagram in circles)
7	Heptad	Septenary, Heptagon, worthy of veneration, number of religion, the number of life, the Motherless Virgin - Minerva, fortune, occasion, custody, control, government, judgment, dreams, voices, sounds, that which leads all things to their end, the mystic nature of man, the number of the law	(heptagram in circles)

Nine To The Ninth Power of Nine

Pythagorean Numerology Table (continued)

#	Name	Description	Symbol
8	Ogdoad	Cube, Air, octahedron, love, counsel, prudence, law, convenience, the little holy number, It derived its form partly from the twisted snakes on the Caduceus of Hermes and partly from the serpentine motion of the celestial bodies; possibly also from the moon's nodes	
9	Ennead	Triple Triangle, Nonagon, the number of humans because of the nine months of embryonic life, Failure, shortcoming, limitless number because there is nothing beyond it but the infinite 10, boundary and limitation because it gathered all numbers within itself, the 9 was looked upon as evil by the Pythagoreans because it was an inverted 6, germinal life because of its close resemblance to the spermatozoon	
10	Decad	Circle, dodecahedron, the system of a universe, assembly point, symbol of earth and heaven, something perfect, the nature of number, the Number of Nature, age, power, faith, necessity, and the power of memory, comprehends all arithmetic and harmonic proportions, heaven and the world, unwearied, tireless,	

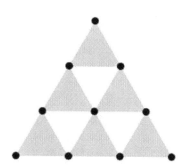

Above: The Tetractys symbol was very important to the Pythagoreans and Kabbalah numerological systems. The Tetractys (or four rows of dots increasing from 1 to 4) was symbolic of the stages of creation.

The Chinese I-Ching book called the Zhouyi is a book used in numerological divination practices that is similar to West African Ifá system. The I-Ching numerological divination book contains 64 sets of 6 lines called hexagrams. Different objects such as Turtle Shells, Coins, Dice, Marbles, Rice, and Bones are used to randomly select which section of the I-Ching is read during the divination process.

#	Trigram Figure	Name	Binary Value	Description
	TAOism I-Ching Divination Numerology			
1	☰	qián	111	Creative, Force, creative, strong, heaven, sky, head, northwest
2	☱	duì	110	the Joyous, Open, tranquil (complete devotion), pleasure, swamp, marsh, mouth, west
3	☲	lí	101	the Clinging, Radiance, clinging, clarity, adaptable, light-giving, dependence, fire, eye, south
4	☳	zhèn	100	the Arousing, Shake, initiative, inciting movement, thunder, foot, east
5	☴	xùn	011	the Gentle, Ground, gentle entrance, penetrating, wind, thigh, southeast
6	☵	kǎn	010	the Abysmal, Gorge, in-motion, dangerous, water, ear, north
7	☶	gèn	001	northeast, completion, resting, stand-still, mountain, hand, northeast
8	☷	kūn	000	the Receptive, Field, receptive, devoted, yielding, earth, belly, southwest

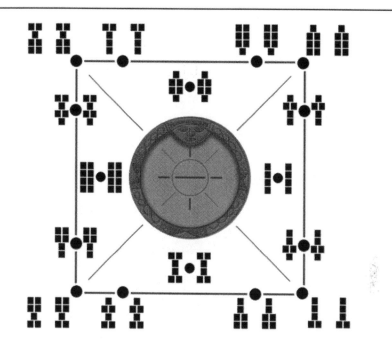

Above: 16 Principle Odu of Ifa from the Yoruba culture in Africa

Above: TAOism's Taijitu (divided circle of Yin and Yang) surrounded by the 8 Trigrams of the I-Ching.

The African Ifá system of the Yoruba people is a system of binary Mathematics that seeks to solve problems and answer questions using a book of knowledge called the **Odú Ifá**. There are 16 Odù Ifá ("Books of Knowledge) each with 16 chapters that when multiplied have 256 chapters believed to reference all situations, circumstances, actions and consequences in life. The Ifá system is considered a combination of Numerology, Bibliomancy, and Geomancy "Divination" using randomization and binary Mathematical methods. The Yoruba Orisha **Orunmila** (also called Orunla, Orunla, FA, Ifa,) is the deity associated with Ifa diviniation system that was said to have been brought to humanity by Orisha **Legba** (also called Papa Legba, Ellegua, Eshu). Orunmila and Eshu are related to the Egyptian Deity **Tehuti** because of their wisdom bringing functions. This is not surprising since the Yoruba people trace their history back to an Ancestor named **Oduduwa** who migrated out of the area of Egypt/Nubia and founded the **Kingdom of Ifa** as the first **Ooni** (or **Oba** meaning Ruler). Performing Ifa divination is called **Idafa** (or **Adafa**) and is done by a "Priest" (Mathematician) called a **Babalawo** (meaning father of secrets). The Babalawo uses various objects (16 Ikin palm nuts, 8 cowry shells on the Opele chain, opon Ifa Divination Tray, etc.) that can randomly generate up to 256 variations of Binary Data. The number randomly generated by "The Ritual" are then marked in powder on the divination tray or drawn on the ground and correspond to one of the 256 chapters of the Odú Ifá to solve whatever problem is being addressed. The similarities between the Asian I-Ching/TAOist system and Ifa system are widely noted. While the I-Ching is dated to about 50 BC, it is predated by the Ifa system which is dated as old as 400 BC, and it is likely that the development of the I-Ching was inspired and motivated by the Ifa system.

Sixteen Principal Odú of Ifá

#	Binary	Name	#	Binary	Name	#	Binary	Name	#	Binary	Name
1	1111	Ogbe	2	0000	Oyẹku	3	0110	Iwori	4	1001	Odi
5	1100	Irosun	6	0011	Owọnrin	7	1000	Ọbara	8	0001	Ọkanran
9	1110	Ogunda	10	0111	Ọsa	11	0100	Ika	12	0010	Oturupọn
13	1011	Otura	14	1101	Irẹtẹ	15	1010	Ọse	16	0101	Ofun

Supreme Mathematic African Ma'at Magic

\multicolumn{8}{c	}{BINARY ORDER OF ODU OF IFA}
Binary Order	Binary Number	Odú	Odú Order	Binary Order	Binary Number	Odú	Odú Order
1	0 0 0 0	Oyẹku	2	9	1 0 0 0	Ọbara	7
2	0 0 0 1	Ọkanran	8	10	1 0 0 1	Odi	4
3	0 0 1 0	Oturupọn	12	11	1 0 1 0	Ọsẹ	15
4	0 0 1 1	Owọnrin	6	12	1 0 1 1	Otura	13
5	0 1 0 0	Ika	11	13	1 1 0 0	Irosun	5
6	0 1 0 1	Ofun	16	14	1 1 0 1	Irẹtẹ	14
7	0 1 1 0	Iwori	3	15	1 1 1 0	Ogunda	9
8	0 1 1 1	Ọsa	10	16	1 1 1 1	Ogbe	1

Left: African Yoruba Orisha deity Orunmila

Left: Vodun Veve sign of Legba

Nine To The Ninth Power of Nine

#	Symbol	Symbol Name	Doctrine	Doctrine Description
			THE 9 TEHUTI DOCRINES	
1		Shen "Called"	Mental	Mind; All; Eternity; The All is in All
2		Thet "Knot"	Correspondence	As Above, So Below; Paradox; Thing; ability to reason from the known to unknown made know; Agreement;
3		Dub "Scarab"	Vibration	motion; rotation; planes of existence; density; master of time and space; Manifestation; Vibration
4		Sekhem "Powers"	Polarity	duality; opposites; different degree; emotional alchemy; ; Rulership; Polarity
5		Ib "Heart"	Rhythm	flow; tides; pendulum swing; ; Emotions; Rhythm
6		Hepet "Embrace"	Cause and Effect	causation; Reality;
7		Sema "Union"	Gender	masculine and feminine; male and female; Reproduction; Gender
8		Shut "Feather"	Growth and Creation	manifestation; apparent change; birth; Order; Growth, Creation
9		Djeneh "Wings"	Breathing	pulsation; in and out; inquisitiveness; Living; Breathing

www.AfricanCreationEnergy.com

What is "Supreme Mathematics"?

"Supreme Mathematics" is a term that is popularly used by members of the "Nation of Gods and Earths" also called "the Five-Percent Nation" which was founded by Clarence 13X (born Clarence Smith and called "Father Allah" by members of the Five-Percent Nation). It widely accepted that Clarence 13X developed the Supreme Mathematics system although some sources state that Clarence 13X got the Supreme Mathematics doctrine from Malcolm X during his time as a member of the Nation of Islam. In either case, "Supreme Mathematics" is Numerological system of assigning qualitative values to the decimal numbers 0 through 9 that worked in conjunction with a system called the "Supreme Alphabet" which served as a form of Isopsephy Numerology to assign values to the meanings of letters of the English alphabet and words spelled in English. The "Supreme Mathematics" numerological system closely resembles Kabbalah Numerology which is not surprising since both systems were influenced by similar monotheistic religious traditions (Judaism and Islam).

Nine To The Ninth Power of Nine

Left: Clarence 13X, Author of the "Supreme Mathematics" Numerology System.

Supreme Alphabet	
A – Allah	**N** – Nation, Now
B – Be or Born	**O** – Cipher
C – See	**P** – Power
D – Divine, Destroy	**Q** – Queen
E – Equality	**R** – Rule or Ruler
F – Father	**S** – Self, Savior
G – God	**T** – Truth or Square
H – He or Her	**U** – You
I – I, Islam	**V** – Victory
J – Justice	**W** – Wisdom
K – King, Kingdom	**X** – Unknown
L – Love or Lord	**Y** – Why
M - Master	**Z** – Zig, Zag, Zig (meaning knowledge, Wisdom and Understanding)

An alternate version of the "Supreme Mathematics" numerology system was developed called "The Twelve Jewels" for the decimal numbers 1 through 12. "The Twelve Jewels" Numerology system is as follows:

The 12 Jewels:
01. Knowledge
02. Wisdom
03. Understanding
04. Freedom
05. Justice
06. Equality
07. God
08. Clothing
09. Shelter
10. Love
11. Peace
12. Happiness

Above: Symbol of the "Nation of Gods and Earths" also called "the Five-Percent Nation" or "The 5-percenters"

"SUPREME MATHEMATICS" NUMEROLOGY

#	Principle	Details
1	Knowledge	knowledge = facts = foundation of all existence = light = sun = original man.
2	Wisdom	wisdom = spoken and active knowledge = water = vital building block of life = original woman = reflection of knowledge = moon (knowledge + the reflection of knowledge=wisdom(1+1=2))
3	Understanding	understanding = show and prove completion of knowledge = clear comprehension = original child = star = love (the bond between man and woman or knowledge and wisdom)
4	Culture or Freedom	culture = way of life = language (wisdom) and customs (ways and actions) freedom = free mind = no restraints
5	Power or Refinement	power = force = creative energy = truth = mathematics = knowledge + culture (1+4=5) refinement = perfect
6	Equality	equality = state or quality of being equal = equal in knowledge wisdom and understanding knowledge + wisdom + understanding (1+2+3=6)
7	God	god = the supreme being = black man who is original = supreme ruler of the universe = everything = sun, moon, and stars (man, woman, and child)
8	Build or Destroy	build = elevate mentality of self and others = addition of positive energy destroy = ruin by allowing negative to outweigh positive
9	Born	born = completion of existence = to manifest from knowledge = completeness in itself
0	Cipher	cipher = a whole = womb = all in existence = circle = (360° degrees = 120° knowledge + 120° wisdom + 120° understanding)

7. SACRED GEOMETRY MATHEMATIC OBJECTS

Mathematical Objects are signs, symbols, insignias, words, and/or phrases used to represent the various abstract concepts and ideas in Mathematics. Examples of mathematical objects include Numbers, Functions, formulas, groups, sets, points, lines, and shapes. The Mathematics of Geometry is needed in order to write or inscribe any Mathematical concept, and so the best and most efficient Geometric symbol for a written Mathematical concept are those symbols that closely and clearly correspond to the concept that they intend to represent.

Since mathematics is the methodological foundation of human understanding, the Geometry or Symbols used to express and represent any Mathematical concept could and should be cherished, revered, and considered sacred. The mathematics and geometry of the blueprint or schematic that enabled the construction of a house that provides safety and security, the electricity to cook food that provides nourishment and the building of a car that provides transportation could all be considered "sacred" since these objects have become almost necessary for everyday life.

When the word "Sacred" is used, it does not mean "sacred" in the religious sense of the word which would involve prayer, meditation, or chanting of some type, but "sacred" in the definition sense of the word as something that should be greatly appreciated and valued. Beyond regular blueprints, schematics, renderings, symbols, graphics, writings, or art work used to represent various concepts and ideas that could be considered "sacred", mathematical analysis into the geometry of Nature has also revealed certain geometric shapes, structures, and patterns that when combined and incorporated with mathematical objects yields the Sacred Geometry of Nature.

Sacred Geometry of Growth and Creation:

Mathematical analysis of Nature has revealed that in order for things to grow balanced without changing shape, they naturally grow in Logarithmic Spirals or Growth Spirals. The Logarithmic or Growth spirals are described mathematically in polar coordinates (r, Θ) by the Exponential power function $r = ae^{b\Theta}$ (where b is called the "growth factor") and when graphed, take on the appearance of the symbol for the quantity Nine (9) in the decimal numbering system. In fact, the symbol for the mathematical constant e (which is approximately 2.718...) also resembles a logarithmic growth spiral. The Greek letter mathematical symbol theta Θ comes from the Egyptian symbol for the sun ☉. Plants, Animals, Weather, Humans, and Galaxies all naturally grow in Logarithmic spirals in order to maintain balance (Ma'at) while growing (Creating).

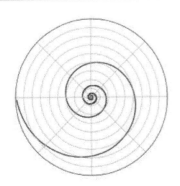
Graph of Logarithmic Growth Spiral

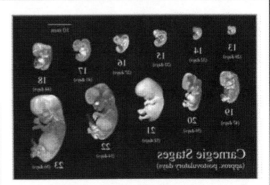
Logarithmic Spiral Growth of Human Embryo

Spiral Growth of Nautilus

Spiral Growth of Weather

Spiral Growth of Galaxies

The ancient Egyptians conceptualized a special Logarithmic Growth spiral that increases for every quarter turn called the "Golden Spiral" where the growth factor was related to the "Golden Ratio". The Golden Ratio between two numbers is when the ratio of the sum of the numbers to the larger number is equal to the ratio of the larger number to the smaller number. The golden ratio is approximately equal to 1.6180339887... and is written in mathematical symbolic notation by the Greek Letter PHI (Φ) which is basically a merged 1 and 0 (or I and O for "Input and Output" in Computer Science), and was derived from the Egyptian Ankh Symbol ☥. The golden ratio and golden spiral were considered

"Sacred" because they were found in Nature and thus the golden ratio and golden spiral were regularly incorporated in Egyptian Architecture, design, art, and blueprints.

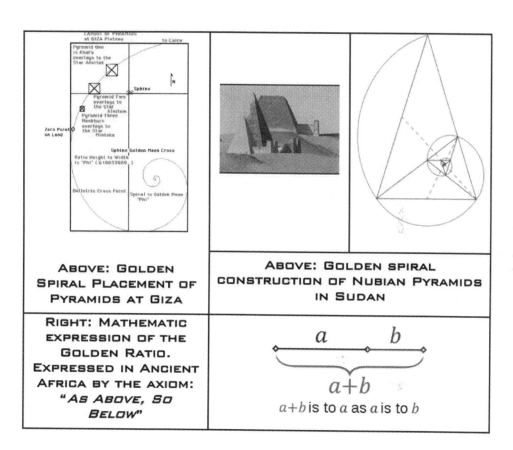

ABOVE: GOLDEN SPIRAL PLACEMENT OF PYRAMIDS AT GIZA

ABOVE: GOLDEN SPIRAL CONSTRUCTION OF NUBIAN PYRAMIDS IN SUDAN

RIGHT: MATHEMATIC EXPRESSION OF THE GOLDEN RATIO. EXPRESSED IN ANCIENT AFRICA BY THE AXIOM: "AS ABOVE, SO BELOW"

$a+b$ is to a as a is to b

Sacred Geometry of Destruction:

Just as the Golden Ratio and Golden Spiral or Logarithmic Growth Spiral can be used to mathematically show Growth and creation, it can also be used to demonstrate the Natural mathematics of destruction. Considering that the spiral can go in two directions: away from the center/origin or towards the center/origin, one spiral can be considered a spiral of construction and one spiral can be considered a spiral of destruction. Mathematics shows the intimate relationship between the seemingly opposite concepts of construction and destruction. The Golden Ratio was developed as a way to geometrically divide a geometric figure into two unequal parts (for example a line or rectangle) such that the ratio of the two smaller parts (offspring) is the same as the ratio of the larger part to the original (parent). This mathematical replication or reproductive process is made possible by the mathematical Golden ratio PHI which comes from the Egyptian Ankh which was called the "Key to Eternal Life" or "perpetual or continued reproduction (**Multiplication by Division**)" as is the case with the Golden Ratio.

Greek Letter PHI, modern Mathematical symbol for the Golden Ratio used for continuous division	African Egyptian Ankh symbol, origin of the Greek letter Phi and the concept of "Eternal Life"

The use of the Golden Ratio to create parts that are similar to the whole is applied mathematically to create Geometric Fractal shapes. The self-similarity property of Natural Fractal Sacred Geometry can be seen in the structure of Solar system compared to the structure of the atom.

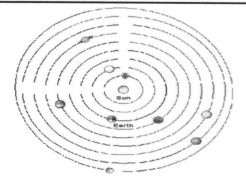

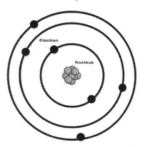

So Below: Sacred Geometric structure of an atom similar to a solar system

As Above: Sacred Geometric Fractal structure of the Solar System is similar to the smaller Atom

  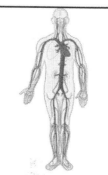

Above: Nature's Fractal Sacred Geometry found in Lightening, Trees, and Human circulatory system

Above: Natural Fractal Sacred Geometry found in Plants

The prevalence and importance of spirals in the "Sacred" geometrical construction of symbols as a reflection of the comprehension of Nature can be found throughout various African mathematical objects. For example, the Akan Adinkra "Gye Nyame" symbol for the Omnipotence and Omnipresence of the One Supreme Being shows two growth spirals depicting the duality of the principles of construction and destruction. The dual principles of "construction" and "destruction" are used in the geometry of the "Gye Nyame" symbol to indicate the "Completeness" or "Oneness" of the Akan deity Nyame to signify that within completeness there is duality. The Geometry of these two Logarithmic spirals resembles the shapes of the numbers 6 and 9. Geometrically, the shapes of 6 and 9 are 180° rotations of one another. In the African Binary numbering system, the Number 9 = 1001 is the binary complement to the number 6 = 0110; hence the reason why these Geometric symbols could be used to denote opposing concepts and ideas.

Nine To The Ninth Power of Nine

Below: Analysis of the Akan Adinkra "Gye Nyame" symbol shows it is comprised of two Logarithmic Growth spirals that resemble the numbers 6 and 9.

Below: Opposite Logarithmic spirals found in the headdress of the Ekpe (Leopard) African Secret society amongst the Ekoi people of West Africa.

Below: Opposing Logarithmic spirals can also be found in the Sacred Geometry of Asia in the TAOist Taijitu symbol (divided circle of Yin and Yang)

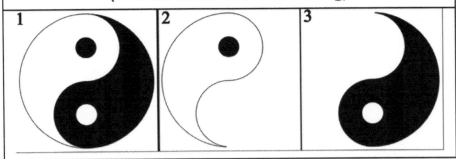

Nine To The Ninth Power of Nine

The two Logarithmic spirals (6 and 9) can be combined geometrically to form a symbol that resembles the number 8 or infinity (∞). The 8 or infinity (∞) geometric symbol is also prevalent in cosmograms and other Sacred Geometry found in Africa and throughout the world. The geometric symbol 8 or ∞ which displays the dual principles of geometric figures 6 and 9 is called the "Vesica Piscis" which means "the bladder of a fish" which alludes to the concept of Primordial waters called Nu or Nun in Egypt and also to the "fish" or "whale" called Nuwn in Hebrew that swallowed Jonah in the Biblical story. The Vesica Piscis Sacred Geometrical shape is the origin of the Venn Diagrams used in Logic, Probability, and statistics. In modern philosophy, the Vesica Piscis shape is used to represent the concept of the Dyad or Duad which is considered the divine mind (nous) and the demiurge or creative force in Nature. In ancient Africa, all creation stories have in common that Nu or Nun was the source from which creation sprang. The shape of the Vesica Piscis is also the shape of the orbit of the **Binary Stars** Sirius A and Sirius B that is significant to the Dogon people of Mali in West Africa.

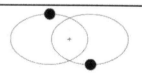

Above: Vesica Piscis shaped orbit of Binary Stars like Sirius A an Sirius B which are significant to the Dogon people of Mali in West Africa

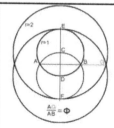

Above: Calculation of Phi in Vesica Piscis shape

Above: The Intersection of the Vesica Piscis shape is the Origin of the Religious symbol of the Fish called "Nuwn" in Hebrew

8. The MATH of Math and The MATH in Math

Mathematics is defined formally as the systematic study of Nature using symbols to obtain solutions through rigorous deductive reasoning. Therefore, "Reasoning" is the fuel that powers the engine of Mathematical methods and thought processes. The process of mathematically describing and writing the Mathematical Reasoning of Mathematical theorems and Mathematical methods in symbols, words, syntax, and notation is called **Meta-mathematics**. The prefix "Meta-" means "beyond", and thus "Meta-Mathematics" means "beyond mathematics" or "beyond the thought process". Of course, beyond the thought process is "the Reasoning process" or the reason for thinking. Unlike "Metaphysics", Metha-mathematic is not considered a pseudoscience or pseudo-math, but rather the foundation and a principle of Mathematics. Meta-mathematics is **"The Mathematics of Mathematics"** (or **"The Math of Math"**) and is the study of Mathematics using Mathematical methods. Meta-mathematics is closely related to the fields of Mathematical Logic and Propositional Calculus in that all of these fields deal with the Mathematic notation of Logic and Reasoning. Whereas Mathematical Logic and Propositional Calculus methods can be applied outside of the field of Mathematics, Meta-mathematics uses these methods for the purpose of describing the supporting reasoning of Mathematical Theorems, Axioms, Propositions, Statements, Concepts, Objects, and Methods.

Meta-mathematics is sometimes considered **"Mathematical Philosophy"** or **"Proof Theory"**. In order to provide a convincing Meta-mathematical description of a Mathematical concept, the Reasoning for the Thought must be presented, which of course would require "Thinking about Thinking" and "Thinking about Thoughts" to provide reasonable Proof to show that the conclusion of the Thought is valid. Meta-Mathematics is the Application of Pure Mathematics not for the purpose of calculation, but rather for the purpose of developing the mathematical methods and theorems that can then be applied to perform a calculation or obtain a result or solution. Meta-mathematics is the thought process that creates thought processes which create actions. In computer programming, Meta-mathematics would be analogous to a Computer code or Program that creates or writes other Computer codes or programs that perform functions.

Mathematics as the systematic quantification, comprehension, and study of Nature has a **Method**, field, or subject area that relates to every aspect of Nature. The various fields of Math, and the Mathematical operations for each Mathematic Method are **"The Mathematics in Mathematics"** (or **the Math in Math**). Mathematical Methods and Operations are the actions, procedures, and processes performed in Mathematics. The word **"Method"** comes from the Latin word "*Methodus*" meaning "way of teaching, traveling, or inquiry" and the word **"Operation"** comes from the Latin word "*Opus*" meaning "work" which is related to the word "Urge" and "Energy" which also mean "work". From the etymological definitions given it can be seen that the Mathematic Methods and Operations that serve as "The Mathematics in Mathematics" are the mental pathways or routes that must be traveled in order for mental energy to be used to workout the solutions to any problems. If Nature is defined scientifically as all Matter or Energy, all Space or Vacuum, and all Time or Existence, then it is clear to which mathematical fields can and are used to study certain aspects of Nature.

Arithmetic, which is a Greek word meaning "Number" or "the art of counting", forms the base of the Science called Mathematics, which is the Science used in all fields of Science and thus is THE SCIENCE in sciences. Arithmetic is the mathematical process of counting and quantification, and thus is applied when studying all of the aspects of Nature. Numbers are the primary mathematical objects used in Arithmetic. A number is a word or symbol use to signify the magnitude or multitude of quantity. Arithmetic is the study of Numeric Mathematical Operations. Numbers are used for counting, measuring, sequence, order, labeling, and coding. There are many different types of numbers including, but not limited to:

- **Natural Numbers** (Whole Numbers) are used for counting and ordering
- **Real Numbers** include both <u>rational</u> and <u>irrational</u> numbers

- **Rational Numbers** are any numbers that can be obtained by dividing two Integers
- **Irrational Numbers** are numbers that cannot be obtained by dividing two Integers
- **Imaginary Numbers** are numbers based on *i* which is symbolic of the value of the square root of -1
- **Complex Numbers** are the combination of Real Numbers and Imaginary Numbers.
- **Positive Numbers** are Real numbers that are greater than zero.
- **Negative Numbers** are Real Numbers that are less than zero. Zero is neither positive or negative (Neutral)
- **Integers** are positive and negative Natural Numbers
- **Ordinal Numbers** are numbers used for establishing rank, order, sequence, or position
- **Cardinal Numbers** are numbers used to measure magnitude and multitude
- **Transcendental Numbers** are non-algebraic numbers like Pi (π)
- **Algebraic Numbers** are complex numbers that are solutions of a non-zero polynomial equation with rational coefficients.
- **Nominal Numbers** (Numerals or Number names) are used for labeling and naming

- **Prime Numbers** are numbers that can only be divided by 1 and itself
- **Odd Numbers** are integers that are divisible by 2 with a remainder
- **Even Numbers** are integers that are divisible by 2 without a remainder

Arithmetic Mathematical Operations on numbers also form the base of all Mathematical Operations and are thus THE OPERATION of operations, and the Order which arithmetic mathematical operations are performed are the Ma'at of Math (Order of Math). The Mathematical methods or operations of Arithmetic are:

- **Addition** is the mathematical process of *putting things together*. (Building)
- **Subtraction** is the inverse or opposite of addition and thus is the mathematical process of *taking things apart*. (Destroying)
- **Multiplication** is the mathematical operation of **scaling** to increase one number by another and is defined as repeated Addition.
- **Division** is the inverse or opposite of multiplication and thus is the mathematical operation of **scaling** to decrease one number by another and is defined as repeated subtraction.

- **Exponentiation** is the Mathematical process of Growth, repeated Multiplication (Repeated scaling over time). Depending on the polarity (negative or positive) of the exponent (also called "Power"), exponentiation can be Exponential Growth or Exponential Decay

Using the African philosophical concept that "To Build is to destroy", the list of Arithmetic operations can be simplified. Addition and subtraction are related because subtraction is just addition with Negative numbers. Multiplication and Division are related because division is just Multiplication with Rational numbers. Therefore, the three basic arithmetic mathematical operations are:

- Addition – Building with Matter
- Multiplication – Scaling in Space
- Exponentiation – Growing over Time

Algebra builds on the concepts from Arithmetic to further provide mathematic methods that enable the comprehension and description of the structure of the various aspects of Nature. Algebra, which comes from an Arabic word meaning "reunion of broken parts", is the Mathematical field that studies the formulas, forms, equations, construction, and structure of Nature by using the Ma'at of Math (the order or laws or rules of Mathematic operations). Algebra is used to study and comprehend all aspects of Nature and is used in all fields and methods of Mathematics.

Geometry along with Trigonometry are fields of Math that provide methods of studying the size, shape, and position of the **Matter** and **Space** aspects of Nature.

Calculus is the Mathematical field that provides methods to investigate the changes in Nature over **Time**. Calculus introduces the mathematic operations of **differentiation** and **integration** which are mathematic methods that enable comprehension and description of the changes in the formulas or equations of **Matter** or **Space** at a point in **Time** (differentiation); and the changes in the formulas of **Matter** in **Space** over a span of **Time** (integration) respectively.

Mathematic **Logic** provides mathematic methods to comprehend and describe the mental processes of Reasoning, and **Probability** provides mathematic methods to comprehend and describe the mental processes of Assumption and Belief.

Arithmetic, Algebra, Geometry, Calculus, Logic, and Probability are the basic broad titles for the numerous Mathematical fields which each sub-divide and combine into various subjects, topics, and areas that provide Mathematic methods for studying the various aspects of Nature.

9. NINE TO THE NINTH POWER OF NINE: 9^{9^9}

In Section 7 it was shown how the geometric symbol for the Quantity nine (9) in the decimal numbering system has the same shape as the Logarithmic Growth Spirals that are found in the "Sacred" geometry of Nature. In Sections 7 and 8 it was shown how Exponentiation is the Mathematical operation of building, growth, and creation.

In this Section, the Mathematical problem "*nine to the ninth power of nine*" (9^{9^9}), which inspired the writing of this book, will

be discussed both the esoterically and exoterically to explain its significance to the act of Creation, as well as to the ancient African concept of Engineering called "Heka" or "Magic". Considering that the symbol for the number nine (9) in the decimal numbering system is a Logarithmic Growth Spiral, and that Exponentiation is the Mathematical Operation of growth, it follows that the mathematic expression "*nine to the ninth power of nine*" (9^{9^9}) is esoteric mathematic symbolism for

Growth and Creation.

Considering that most computing devices at the time of this writing are not pre-programmed to be able to process the calculation of the mathematic expression "*nine to the ninth power of nine"* (9^{9^9}), it would require someone knowledgeable

enough to innovate or modify the existing computing devices, or engineer, invent, or create a new computing device in order to be able to exoterically calculate the numeric quantity of the mathematic expression *"nine to the ninth power of nine"* (9^{9^9}),

Because the African Binary Base-2 numbering system is the foundation upon which all computations in modern computers and electronic devices are built, then application of the knowledge of the African Binary base-2 numbering system provides the ability to creatively modify the existing framework of modern computers to generate a numerical solution to the mathematic expression 9^{9^9}.

There are many computational algorithms that can be used to calculate the numerical solution to the mathematic expression 9^{9^9}. The particular algorithm that was used that requires

knowledge of the African Binary Base-2 numbering system is called **"Exponentiation by Squaring"** and is an algorithm that has other applications in the field of **Cryptography** which is the mathematical study of encoding or **hiding** and **decoding** or uncovering information. The "Exponentiation by Squaring" algorithm provides the ability to break down the mathematic expression 9^{9^9} as follows:

Nine To The Ninth Power of Nine

$9^{9^9} = 9387420489$

$9387420489 = 9101110001011110010001010010001_{bin}$

$= 9^{2^{28}} * 1 * 9^{2^{26}} * 9^{2^{25}} * 9^{2^{24}} * 1 * 1 * 1 * 9^{2^{20}} * 1 * 9^{2^{18}} * 9^{2^{17}} * 9^{2^{16}} * 9^{2^{15}} * 1 * 1 * 9^{2^{12}} * 1 * 1 * 1 * 9^{2^{8}} * 1 * 9^{2^{6}} * 1 * 1 * 9^{2^{3}} * 1 * 1 * 9^{2^{0}}$

$= 9^{268435456} * 9^{67108864} * 9^{33554432} * 9^{16777216} * 9^{1048576} * 9^{262144} * 9^{131072} * 9^{65536} * 9^{32768} * 9^{4096} * 9^{256} * 9^{64} * 9^{8} * 9^{1}$

$= (((((((((((((((((((((((((((9^2)^2)9^2)9^2)9^2)^2)^2)9^2)^2)9^2)9^2)9^2)9^2)^2)^2)9^2)^2)^2)9^2)^2 \quad 9^2)^2)^2 \quad 9^2)^2)^2)9^2)$

Arithmetic and Algebraic mathematic operations can then be easily applied on modern computers to generate a numerical result. The numerical value of the solution to the mathematic expression 9^{9^9} is the largest quantity that can be expressed using only three digits in the decimal numbering system. In the hexadecimal numbering system, F^{F^F}, which equals $15^{15^{15}}$ in the decimal numbering system, would be the largest quantity expressible using only three hexadecimal digits. In the Vegesimal numbering system, the expression for $19^{19^{19}}$ would be the largest quantity expressible using only three Vegesimal digits. The numerical value solution to the mathematic expression 9^{9^9} has 369,693,100 digits. There are 4.5 letters in the average English word, so to display every digit of the numerical value solution to 9^{9^9} would be equivalent to approximately 82,154,022 Words. An "Epic Super-novel" (very large book) has around 120,000 words, so to display the every digit of the numerical value solution to 9^{9^9} would be equivalent to writing 685 "Epic Super-Novels". To display every digit of the numerical value solution to the mathematic expression 9^{9^9} would literally "grow" or "create" a Library. For this reason, we use "Scientific Notation" to display an abbreviated (approximate and rounded to significant digits) numerical value solution to the mathematic expression 9^{9^9}.

$9^{9^9} \approx$ 4.28124773175747048036987115930563521339055
4822414435141747537230535238874717350483531 9366
5299432033375060417533647631000780326139047 3386
0832080206037470612809165574132086446019861 9996
1452031052442858148959811514719493517677965 5930
2159393385015023969426231052967581656943333 4631
4749255384948593377812087624957216504195220 6018
0457130151786405101459407904194866332733667 1830
6544107602348236334279338847344917149071392 8387
6367034707332216158426388470264465058580355 8248
2311577827786618011472099436290690473438366 4886
6469502338173533149328881151761248597120153 3575
6443987605995621733954850395053696554453295 5477
6218333817990375374298660361754107669609047 1833

Nine To The Ninth Power of Nine

9923933145342547022698333065128258703520736236343330809161939992399143539951742626192250444914889355346296338764247108036190948328339353383268116816840967521737160227124038642410944863124167336163160258473857727316993387556729418877537923876279151815197169574861839692092170993607802644764408395926434454851800780948395933285398270164750251095376500141286928505876878853669407158547929415017596380093739697517984175797277906741998154544863852633240819143167556841941071017167324488713374402297730006005442460032552018028477336157137560860266753519143348470953542806215894313422062980873500411731911584053164106252625074754678272488405835202922863455599567929527725263396995135064858726675736711246852441322863857060434188996170746314452550800737145968257096891089328244493 64 × 10^{369,693,099}

The symbol for the quantity nine (9) in the modern decimal numbering system is a logarithmic growth spiral that resembles the symbol for the quantity 100 (◉) in the ancient African Egyptian numbering system. The numeric value of the mathematic expression ◉◉ would be equivalent to $100^{100^{100}}$. In either case, whether we are considering the numeric value of the mathematic expression

Supreme Mathematic African Ma'at Magic

$9^{9^9} \approx 4.2812 \times 10^{369,693,099}$ or we are considering the numeric value of the mathematic expression ೲ = $100^{100^{100}}$, Both of these quantities would be considered "Infinity" (a type of Infinity not "the Infinity of Infinities") and would be expressed in the Ancient African Egyptian numbering system as **Heh** 𓁨 . Heh and his female counterpart Hehet were African Egyptian deities that represented the concepts of Infinity, Eternity, and endlessness.

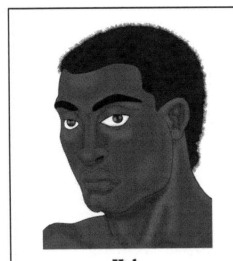

Heh — **Hehet**

Above: Heh and Hehet, African Egyptian deities representing the concepts of Endlessness (space), Infinity (matter), and Eternity (time)

Nine To The Ninth Power of Nine

In the Ancient African Egyptian numbering system, when the solution to the Mathematic expression for either 9^{9^9} = ![], or ![] = ![], is written out vertically for Heh and Hehet, the resulting geometric figure forms the Egyptian Hieroglyphic for **Heka** or "Magic" ![].

Above: In Egyptian numbers and Hieroglyphs, the solution to the Mathematic expression 9^{9^9} forms the Egyptian Hieroglyphic for Heka or "Magic"

The word "Heka" is translated as "Magic" and was associated with the concepts of "power, authority, influence, or control". In Ancient Africa, Heka was deified and depicted as a combination of all of the deities and was used to represent the concept of working with, manipulating, weaving, forging, forming, transforming, building, sculpting, molding, and creating with the chemicals, elements, components, and various forms of matter in Nature. Heka is considered "Magic", but is more closely and reasonably related to the modern concept of **Engineering**. Using Heka to develop various "potions" or tonics to treat certain conditions could be considered a form of Chemical Engineering. Using Heka to create and develop various tools, structures, or other resources could be considered a form of Mechanical Engineering, Architectural Engineering, or Electrical Engineering. Even, the ancient African activity of selectively breeding only with certain family bloodlines could be considered using Heka as a form of Genetic Engineering.

Nine To The Ninth Power of Nine

In the Ancient African system, Heka along with the concept called **Hu** (divine creative utterance, logic, reason, purpose, and intention) and the concept called **Sia** (insight, comprehension, and understanding) were the three principles or concepts that the Creator deity used to create. In modern terms, the three principles or concepts that are used to create, invent, and develop things are Science, Mathematics, and Technology. Engineers acquire and apply knowledge to find solutions and create and invent things to solve problems. The word "Heka" literally meant "to activate the Energy inside one's self". In the Egyptian system, a practitioner of Heka or Engineering had to activate an internal creative energy called **Sekhem**. **Mathematics** is the mental tool that provides the **insight** and comprehension which enables the creative energy of African people to engineer, invent, innovate, and create anything that is needed or desired. The creative energy of African people is **African Creation Energy**.

Q.E.D. ∎

0. ABOUT THE AUTHOR: WHAT IS AFRICAN CREATION ENERGY?

The word "**Physics**" is a Greek word meaning "**Nature**" and has been used to refer to the scientific study of Space, Matter, and Time. In the field of Science called Physics, "**Energy**" is defined as simply "The Ability to do **work**" or the amount of **Work** that can be done by a **Force**. In the field of Science called Physics, a "Force" is defined as any thing that causes the **Change of Position** and hence suggests **Movement**. Only the field of Science called Physics (Nature) in the area of **Thermodynamics** (which studies the movement and transformation of Energy) can there be found any Scientific Laws (that which is not assumed but rather known and accepted as an undisputable fact). One of the Laws of Physics (Nature) is called "**Conservation of Energy**". The Conservation of Energy Law of Nature (Physics) states that **Energy cannot be created** in the sense that it comes from Nowhere and Nothing into existence, and **Energy cannot be destroyed** in the sense that it ever ceases to exist, but rather **Energy can Transform or change from one form to another**. Energy has many forms including Potential Energy, Kinetic Energy, Atomic Energy, Heat Energy, Electromagnetic Energy, Chemical Energy, Gravitational Energy, Sound Energy, and all phases of Matter. Everything in existence is made up of some form of energy. In Nature (Physics), **Power** is scientifically defined as the rate or amount of time it takes for **work** to be done or

energy to be converted. In Nature (Physics), **Radiation** is scientifically defined as any process by which the energy emitted by one object travels through a medium and is eventually absorbed by another object. In the field of Science called Physics (Nature), a **Black Body** is scientifically defined as an ideal object that absorbs all electromagnetic radiation (Light Energy) that it receives.

Creation refers to the act or process of causing something new or novel (nova) to exist or come into being. As it follows from the discussion about Energy, since everything in Existence is composed of Energy, and Energy can only change from one form to another, then Creation and Creating does not mean bringing about something new into existence from Nothing and Nowhere, but rather Transforming or changing the form of one thing, or combining the forms of several things, into the desired new thing. Creation of anything in any form is a gradual process of **Growth** and change over time in which a metaphorical or literal seed, nut, kernel, or node grows and transforms into another form, figure or structure. A new creation is always the initiation or beginning of one thing and the termination or ending of another thing. Growth and Decay are mathematical concepts which indicate the increase or decrease of a quantity, magnitude, or multitude over time respectively. Creativity is the level or degree of creative mental ability, and a creation always exists as a thought, idea, or concept in the mind of its creator before manifesting in the physical world. Creation occurs through Creativity, and one of the most important Creativity techniques is "Problem Solving". Problem Solving occurs when it is desired to go from one state to another state (change of position), and therefore from the scientific definitions given, "Problem Solving" and **Creativity** are both a mental **Force** which can be measured as **Energy**.

From the scientific definitions given, it follows that **African Creation Energy** can scientifically be defined as the Work, Effort, Endeavors, and Activities of Black or African people that cause a movement or change. African Creation Energy is The Energy, Power, and Force that created African people and that African people in turn use to Create. Since African people are the Original people on the planet Earth, it follows from thermodynamics that the Creation Energy of African people is the closest creation Energy of all the people on the Planet to the **Original Creative Energies** that created the Planets, stars, and the Universe. **African Creation Energy** is **Black Power** in the scientific sense of the word "Power", and this book **Radiates** African Creation Energy to be absorbed by the **Black Body**.

Amongst the **Yoruba** people in **Nigeria**, **African Creation Energy** is called **"ASHE"** which means "the power to make things happen" and the various forms of African Creation Energy are personified in the Yoruba "Orisha" (Ori-Ashe) deities. Amongst the **Akan** people in Ghana in the Twi language, **African Creation Energy** is called **"TUMI"** – the web of <u>energy</u> and <u>power</u> that exists throughout space and all of creation which was woven and designed by the Akan deity of wisdom Ananse Kokuroko. In the Congo, **African Creation Energy** is called **"DIKENGA"** which refers to the <u>energy</u> of the universe and the <u>force</u> of all existence and creation, and the thermodynamic process of the transformation of energy is depicted in the Congo cosmogram called the "Yowa". In Ancient Africa, Egypt, Nubia, KMT, Kush, Tamare, *etc. et al*, **African Creation Energy** was called **Sekhem** which was an <u>energy</u> that individuals used to control the elements of nature to create anything desired. In books written by Amunnub-Reakh-Ptah, the original creative <u>force</u> of African people is called **Nuwaupu**.

It is the view in the Western world that *"Those who can, <u>DO;</u> and those who CANNOT, teach"*. This statement is used to imply that people who are able to Use knowledge once it is acquired put it into practice and those who do not know how to use the knowledge or do not have the ability to use the knowledge can only "Teach" or relay the knowledge and information on to someone else. This western view about knowledge, information, and Teachers motivates western governments to pay Teachers the lowest salaries and in turn, forces individuals who do have the "Know How" to put Knowledge into practice, out of the classroom; which only further mentally cripples the future generations. **African Creation Energy** is about **CONSTRUCTING** (DOING, TECH) and **INSTRUCTING** (TEACH) – doing what is necessary to apply all knowledge (Technology) acquirable and practical for the well being of African people NOW, and teaching, instructing, and relaying the Knowledge and the Know How to ensure the well being of African people LATER. With that being stated, the physical writer of this book - who is African by blood and lineage, a descendant of the Balanta and Djola tribes in present day Guinea-Bissau (Ghana-Bassa) in West Africa, a descendant of the Ancient Napatan, Merotic, Kushite Empire, and a Scientist, Engineer, Mathematician, Problem Solver, Analyst, Synthesizer, Artist, Craftsman, and Technologist by education, profession, and nature - has set out to develop, engineer, invent, formulate, build, construct, and create several Technologies (Applications of Knowledge) for the well being of African people world wide and has attempted to relay and teach the information that motivated and inspired the development of those technologies in a three part introductory educational series which collectively have been entitled "The African Liberation Science, Math, and Technology Project" **(The African Liberation S.M.A.T. Project)**. The three books that are part of African creation Energy's "African Liberation S.M.A.T. project" are:

1. **SCIENCE** (Knowledge/Information): "The SCIENCE of Sciences, and The SCIENCE in Sciences"
2. **MATHEMATICS** (Understanding/Comprehension): "Supreme Mathematic African Ma'at Magic"
3. **TECHNOLOGY** (Wisdom/Application): "P.T.A.H. Technology: Engineering Applications of African Science"

Nine To The Ninth Power of Nine

The purpose of African Creation Energy's "African Liberation S.M.A.T. project" is to free the minds, energies, and bodies of African people from mental captivity and physical reliance and dependence on inventions and technologies that were not developed or created by, of, and for African people. In the western world (America and Europe) there are certain technologies such as indoor plumbing, electricity, computers, internet, heating and air conditioning, television, radio, and automobiles that have become so commonplace in the lives of people that it is difficult for some people to imagine life without these technologies. Conversations with the most "Afro-Centric" or "conscious" Black person in the Western World reveal that one of the primary reasons why they prefer to live in the Western World rather than in Africa is for the technologies that they have become used to in America or Europe. A significant amount, if not all, of the inventions needed for survival and well being that African people in Africa and the Diaspora use globally have been invented and developed by individuals other than Africans. Therefore, it is observed that the majority of Creative Energies of African people in Africa and the Diaspora are directed toward purely aesthetic and entertainment creative expressions and endeavors such as music, art, and fashion. The creativity of Black or African people has dominated and influenced the world in the areas of art, entertainment, culture, and fashion, and the degree and level of African Creation Energy can be seen and examined in the degree of Creativity that is exhibited by African people in the areas of music, art, culture, entertainment, and fashion globally. However, Creative expressions such as music, art, and fashion <u>solely</u> for the purpose of aesthetics and entertainment should only come after Creative Energy is used for the purpose of creating, developing and inventing everything needed for survival and well being. Therefore, the solution to Liberate African people from the spellbinding reliance and dependency on Western, European, or Non-African technology is to engineer, invent, and Create African technology. The word "Technology" means "applied knowledge", therefore, to be dependant on the "Technology" of another is to be dependant on someone else to think for you! Technology or "applied knowledge" in the broadest sense, is the entity that is the most spellbinding and captivating for African people by non-Africans. Religion is a technology or invention that applies amongst other sciences, the sciences of Psychology and Sociology which spellbinds and captivates the minds of African people. Money is a technology or invention that applies amongst other sciences, the Science of Economics, and spellbinds and removes African people away from real Natural resources. Technology in the form of Inventions applies science to make life easier and spellbinds the Minds, bodies, and energies of African people making African people dependant on the thinking and creativity of other people. Religion, Money, and Technology are the three most spellbinding and captivating forces that African people must be Liberated from in order to advance, and all of these entities fall under the umbrella of "Technology" (Applied Knowledge).

Therefore, it is the goal of African Creation Energy to Liberate African people from the "Spell of Technology".

It was Technology, either in the form of weapons, doctrines, or other inventions, that enabled Europeans to colonize, enslave, conquer, perpetuate, and enforce their rule over African people and the world. Anything that threatens the survival, well-being, and free expression of African culture and concepts is indeed oppressive and requisite for Liberation. Liberation from oppressive entities may require physical conflict, but until African people create African technology that is competitive with, or better than those of our oppressors, engaging in a physical campaign for liberation is understandable from an emotional standpoint, but premature in terms of being sincerely practical resulting in a positive outcome. It is unintelligent to go up against tanks and machine guns with literal or metaphorical sticks and rocks, this action will only lead to slaughter. It is likewise unintelligent to go to war using weapons and technology that was invented by the very foe that is being fought. The metaphorical SPEAR AND SHIELD (offensive and defensive) technology of Africans must be advanced, upgraded, and progressed to ensure the salvation of African culture and well being after Liberation. The Asian writer Sun Tzu wrote about "The Art of War", however, having the ability to create the necessary technology to ensure that African people are not colonized, enslaved, exploited, or taken advantage of, and ensuring that traditional African culture is preserved is "The Science of War". Technology in the forms of weapons (such as guns, bombs, missiles, lasers, etc.) that threaten the survival, existence, and well-being of Africans are the fear inducing, supporting, and driving forces of the success of other technologies such as Law systems, Religious systems, and economic systems because without the "enforcer" technologies in the form of weapons, African people would immediately see the flaws in the Law, Religious, and Economic systems and would quickly abandon them. In addition to weapon technology as the physical active "Enforcer" of the spell-binding and captivating mental technologies such as Law, Religious, and Economic systems, survival technologies that make day-to-day life easier and simpler (such as electricity, plumbing, automobiles, phones, etc.) are the dependency creating passive enforcers and active encouragers of African people to willingly partake and participate in non-African Culture, Law, Religious, and Economic systems.

The importance of Liberating African people from the dependency and reliance on Non-African technology is because an invention or creation is just like an offspring; it comes from its creator and therefore has its creator's Nature and likeness. A "Brainchild" in the form of a concept, idea, invention, or creation is just like an actual child in that certain traits and characteristics of the creator are passed on to the creation. Therefore, a person, place, or thing has the nature and likeness of whoever created it. If the creator or inventor of a piece of technology needed for survival has a Natural disposition that is not in favor of the survival and well-being of

African people, then that piece of survival technology is actually detrimental to the survival and well-being of African people. In the western world there have been many Scientist, inventors, engineers, doctors, etc of African descent that are responsible for many technological advances in the western world. Lewis Latimer, George Washington Carver, Benjamin Banneker, Elijah McCoy, Garrett Morgan, Granville T. Woods, Philip Emeagwali, Patricia Bath, Marie Van Brittan Brown, Jane Cooke Wright, Louis Tompkins Wright, etc. were African people that contributed significantly to the revolutionary, technological, and industrial advances that made the United States into the world power that it is today. In order for Africa to advance significantly in the same manner, African people worldwide must contribute their respective creative energies to the development and advancement of Africa. There have been many Afro-centric (African centered) revolutionary **movements** (forces or energies) and organizations for a variety of causes that have emerged over the years. Economic movements such as "Black Wall-street," "Ujamaa - Cooperative Economics" and others have been aimed at empowering African people financially and economically. Movements such as "The Black Panther Party", "Pan-Africanism", "Uhuru", and "African Socialism" have been focused on empowering African people politically; and numerous organizations have been developed to promote, teach, and empower African spirituality, religiously, historically, and culturally. The African Creation Energy **Movement** (or Force) is one that not only **empowers** the **Black Body** of African people creatively, scientifically and technologically, but also Creates **Power** from African Science and Technology and seeks to be the catalyst in the **African Technological and Industrial Revolution**. If African people need something, then we should possess the Creativity, Knowledge, and Will to create it; in Africa this is referred to as "**Call and Response**" – the basis of problem solving and a fundamental aspect of Creativity. As long as African people rely on others to do our Thinking for us is as long as the problems of African people will persist. African Creation Energy provides the creative insight and motivation as well as the correct scientific knowledge to respond to the calls, answer the questions, solve the problems, and create solutions for African people.

The phrase "Reinventing the wheel" is a metaphorical phrase used to mean the duplication, re-constructing, or re-inventing of a basic idea, object, method, technology, or other invention that is known to already exist and be in use. The metaphor "Reinventing the wheel" is based on the fact that the wheel is a prototype, basic, simple, and fundamental human invention that is the corner stone of other technologies and not known to have any operational flaws. Therefore, the metaphor "Reinventing the wheel" attempts to symbolically express a creative endeavor that would seem pointless and add no value to the object, and would be a waste of time, diverting the inventor's energy from possibly more important problems which his or her energy could be put to use. However, African Creation Energy reiterates, "A creation or invention (person, place, or thing) has the Nature of whomever or whatever created or invented it". Therefore, if the metaphorical wheel was created or invented by someone or something who's Nature is not in favor of the survival and well-being of us African people, then us African people will be and are using tools and inventions of our own destruction. For this reason, African Creation Energy must Re-invent the metaphorical wheel whenever necessary. Of course, any duplication, reinvention, or reproduction of a creation or invention that already exists must be done within Reason and in accordance with Nature for the survival and well being of African people everywhere.

At this point it is reiterated, the word "Technology" means "Applied Knowledge" and therefore does not just refer to electronics, tools, gadgets, and/or gizmos; but rather to any and all persons, places, and things that can be created, developed, formulated, manufactured, engineered, or invented by applying and using knowledge. This book is written by African Creation Energy for African Creation Energy to develop not just "Free Energy" in the sense that you can light your house and drive your car for free, or no continuous monetary cost, but **"Liberation Energy"** in the sense that you are mentally liberated to know how to do what ever is needed to accomplish any desired goal or create any solution. African Creation Energy is Real Black Power, not just a cliché, but Creative Black Power that can light your house, run your car, teach your children, and guide your government.

Supreme Mathematic African Ma'at Magic

"The Ethiopian Race is not only physically captive to MANKIND but also captive to his CULTURES and INVENTIONS, and this means Wooly-Haired People are mental captives to adverse forces as well as physical. Mental captivity is the worst kind of captivity because it means that the captive is mentally dependent on the captor. Mankind is all peoples with straight hair by Nature… If we need something we must work and get it or **create it***. So long as we remain mentally dependent on other races to do our thinking and plan our progress, that is how long we will remain in the GHETTOES and SLUMS and the POVERTY and DISEASES that they produce…The REASONABLE and PRACTICAL worldwide solution is MENTAL LIBERATION which produces* **ETHIOPIAN CREATIVENESS** *(the ability to create everything we need) and ETHIOPIAN INDEPENDENCE (the ability to think independent and act independently of other races)… A person, place, or thing is alien to us Ethiopians when it is not of and for us by Nature. A person, place, or thing is in the Nature of whoever created it."*

-The Nine Ball Liberation Information

Nine To The Ninth Power of Nine

REFERENCES:
1. "Africa Counts: Number and Pattern in African Cultures" by Claudia Zaslavsky
2. Africa: The True Cradle of Mathematical Sciences
   http://www.africamaat.com/Africa-The-true-cradle-of
3. "African Fractals: Modern Computing and Indigenous Design" by Ron Eglash
4. How Menstruation Created Mathematics
   http://www.tacomacc.edu/home/jkellerm/Papers/Menses/Menses.htm
5. "Introduction to The Nature of Nature books 1 and 2" by Afroo Oonoo
6. Mathematicians of the African Diaspora
   http://www.math.buffalo.edu/mad/Ancient-Africa/ishango.html
7. Rediscover Science http://www.recoveredscience.com/index.htm
8. The African roots of mathematics. Teacher's resource guide by Deborah Lela Moore
9. "The Nine Ball Liberation Information Count I, II, III, IV"
   written by Wu Nupu, Asu Nupu, and Naba Nupu
0. "What is Nuwaupu?" by Amun nebu Re Akh Tar

RESOURCES:
- "An Introduction to the History of Mathematics" by Howard Eves, Holt, Rinehart & Winston
- "A Young Genius in Old Egypt" by Beatrice Lumpkin
- "Black Mathematicians and Their Work" by Virginia Newell
- "Blacks in Science: Ancient and Modern" by Ivan Van Sertima
- "Fascinating Fibonaccis: Mystery of Magic in Numbers" by Trudi H. Garland
- "Golden Legacy" by Baylor Publishing Co. and Community Enterprise, Inc., Seattle, WA, 1983
- "History of Mathematics" by Arthur Gittleman
- "Mathematics in the Making" by Lancelot Hogben
- "Mathematics in the Time of the Pharaohs" by Richard Gillings
- "Secrets of the Great Pyramid" by Peter Tompkins, Harper & Row
- "Stolen Legacy: Greek Philosophy is Stolen Egyptian Philosophy" by George G.M. James
- "The Rhind Mathematical Papyrus" by Arnold Chase

PHOTO CREDITS:
1. Page 8, Four figures for Boolean algebra article, 7 July 2007, Author=Vaughan Pratt
2. Page 32, Quartz Clock photo: Photo taken with a Canon PowerShot A300® 3.2mp 5mm by Felipe Micaroni
3. Page 32, Ishango bone from two different points of view. Museum of Natural Sciences, Brussels
4. Page 35, Moscow Mathematical Papyrus photo: Nederlands: Moskou-papyrus (probleem 14) Date: 22 March 2009(2009-03-22); Author: Quatrostein
5. Page 35, Reisner Papyrus photo by Carlos Avila Gonzalez from the UC Berkeley Bancroft Library
6. Page 36, Rhind Mathematical Papyrus photo by Paul James Cowie (Pjamescowie) from British Museum EA 10057 Department of Ancient Egypt and Sudan, Acquired by the Scottish lawyer A.H. Rhind during his sojourn in Thebes in the 1850s.
7. Page 51, Mayan Numbers, Description=Grafisch vereinfachtes Zahlensystem der Mayas, Date= 2006-11-25, Author=Bryan Derksen
8. Page 56, Black Hairstyles, www.hairstyleslibrary.com
9. Page 104, Carnegie Stages of Human Development, From "The Multi-Dimensional Human Embryo," University of Michigan
10. Page 110, Ekoi skin covered head dress, Photo by Ukabia

Made in the USA
Columbia, SC
25 November 2021

49774240R00088